CLIMATE CHANGE AND TOURISM

The contribution of tourism to climate change, and the likely consequences of climate change for key tourist destinations, has been well reported and discussed. Yet, there is a lack of evidence-based systematic practical advice as to how the tourism industry should respond to the challenge of climate change. Building on a sound conceptual understanding of the links between climate change and tourism, this book shows how the tourism sector might best respond. It not only focuses on the roles of supportive policies and institutions in ensuring a strong 'enabling environment' for practical responses, but also on the practical responses themselves.

This practical approach is presented through a large number of case studies and examples which illustrate how policy and industry initiatives have been implemented in tourism, and if or why they were successful. The majority of examples come from places such as the Caribbean, Spain, the Maldives, Nepal, and the UK, as well as Australia, New Zealand and other parts of the Pacific. The examples are presented within an overall framework that facilitates the translation of adaptation and mitigation policies into practice.

This book offers the tourism industry, students and academics the opportunity to advance from the earlier, more conceptual texts on tourism and climate change by taking a much more practical approach. Its global coverage, through the use of international case studies, fosters a cross-fertilization of ideas and initiatives. This text provides a detailed analysis of best practices in the face of climate change, across countries and geographically diverse tourist destinations and operations.

Susanne Becken is a Professor of Sustainable Tourism at Griffin University, Australia, and an Adjunct Professor at Lincoln University, New Zealand. She is also on the editorial boards of *Annals of Tourism Research*, the *Journal of Sustainable Tourism*, the *Journal of Policy Research in Tourism, Leisure and Events*, and the *Tourism Review*.

John E. Hay is owner and director of JEH & Associates. He is also Visiting Professor at Ibaraki University, Japan, and an Adjunct Professor at both Lincoln University in New Zealand and the University of the South Pacific; and a Lead Author for the Intergovernmental Panel on Climate Change (IPCC).

Tourism Environment and Development Series
Series Editor: Richard Sharpley
School of Sport, Tourism & The Outdoors, University of Central Lancashire, UK

Tourism, Development and the Environment
Beyond sustainability?
Richard Sharpley

Tourism and Poverty Reduction
Pathways to prosperity
Jonathan Mitchell and Caroline Ashley

Slow Travel and Tourism
Janet Dickinson and Les Lumsdon

Sustainable Tourism in Island Destinations
Sonya Graci and Rachel Dodds

Climate Change and Tourism
From policy to practice
Susanne Becken and John E. Hay

Please contact the Series Editor to discuss new proposals at: rajsharpley@uclan.ac.uk

CLIMATE CHANGE AND TOURISM

From policy to practice

Susanne Becken
and John E. Hay

Routledge
Taylor & Francis Group
LONDON AND NEW YORK

First published 2012
by Routledge
4 Park Square, Milton Park, Abingdon, Oxon OX14 4RN

Simultaneously published in the USA and Canada
by Routledge
605 Third Avenue, New York, NY 10017

Routledge is an imprint of the Taylor & Francis Group, an informa business

British Library Cataloguing in Publication Data
A catalogue record for this book is available from the British Library

Library of Congress Cataloging in Publication Data
A catalog record has been requested for this book

ISBN: 978-1-84971-475-4 (hbk)
ISBN: 978-1-84971-476-1 (pbk)
ISBN: 978-0-203-12896-1 (ebk)

Typeset in Bembo
by Cenveo Publisher Services

John E. Hay dedicates this book to his wife, Eleni, who has joined him on a journey few can dream of, let alone experience.

CONTENTS

List of plates, figures and tables x
List of acronyms xvi

1 Introduction 1
1.1 Tourism and climate change 1
1.2 Rationale and structure of the book 4

2 Recent Advances in Knowledge and Understanding 8
2.1 Climate change science 8
2.2 The climate change debate 11
2.3 Policy for climate change and tourism 12
2.4 Measuring greenhouse gas emissions by tourism 15
2.5 Supply-side responses to climate change 18
2.6 The environmental behaviour of tourists 20
2.7 Tourist perceptions of climate and weather 23
2.8 Local impacts and destination-level adaptation 25
2.9 Future-oriented analyses 28
2.10 Conclusion 30

3 Adaptation Policy, Planning and Institutions 31
3.1 Introduction 31
3.2 Ensuring a strong enabling environment 33
3.3 Incorporating managing climate risks into national policies and plans 35

3.4 Adaptation policy making and planning at the sub-national and business levels 41
3.5 Adaptation policy making and planning at multi-country and international levels 50
3.6 Conclusion 52

4. Adaptation Practice 55
4.1 Introduction 55
4.2 Incentive to travel, and appeal of potential destinations 57
4.3 Transport infrastructure and services used by tourists 64
4.4 Managing tourism's resource base 72
4.5 Managing tourist satisfaction, health and safety 83
4.6 Managing the sustainability of facilities, operations and destinations 95
4.7 Conclusion 101

5. Mitigation Policy 103
5.1 Introduction 103
5.2 Business level mitigation policies 105
5.3 Local mitigation policies 112
5.4 National mitigation policies 116
5.5 Regional level mitigation policy 124
5.6 International mitigation policy for tourism 130
5.7 Conclusion 137

6. Mitigation Practice 139
6.1 Introduction 139
6.2 Destination appeal 141
6.3 Transportation 149
6.4 Mitigation at facilities and in operations 163
6.5 Tourist behaviour and mitigation 172
6.6 Conclusion 177

7. Assessing Performance, and Learning 180
7.1 Introduction 180
7.2 Assessing the performance of mitigation policies and plans for the tourism sector 182
7.3 Assessing the performance of adaptation policies and plans for the tourism sector 191
7.4 Maladaptation in the tourism sector 198
7.5 Knowledge sharing and learning 199
7.6 Conclusion 202

8. Integrated Responses to Climate Change 203
8.1 Introduction 203
8.2 Climate change mitigation and adaptation 204
8.3 Sustainable development and climate change response 212
8.4 Disaster risk reduction and adaptation 223
8.5 'Peak oil' and climate change 229

9. Conclusion 235

Notes 240
References 242
Geographical Index 266
General Index 268

PLATES, FIGURES AND TABLES

Plates

3.1 Information for tourists on road conditions after a heavy snowfall event in Wanaka, New Zealand. 35
3.2 Tourist beach-scene in Ocho Rios, Jamaica, on a windy day. 38
3.3 Damage caused to the Barcelo Maya Resort, Quintara Roo, Mexico, as a result of Hurricane Wilma. 39
3.4 Australia's Gold Coast, the recently announced host city for the 2018 Commonwealth Games. 43
3.5 A (former) restaurant destroyed by Category 5 Hurricane Wilma in October 2005, Cozumel, Mexico. 47
3.6 Sagarmatha National Park (Mt Everest) seen from the Tibet side. Increased melting potentially affects climbing and other tourist operations. 52
4.1 Climate change is expected to result in intolerably hot conditions for mid-summer sunbathers, such as these on the coast of Croatia. 60
4.2 Rail line subject to high levels of leaf fall and resulting low adhesion. 68
4.3 and 4.4 Coastal tourism is important to the economy and livelihoods, but also vulnerable to climate change, whether in part of a large country, such as Santa Monica, United States (top) or a small island, such as Aitutaki, Cook Islands (bottom). 74
4.5 Traditional tourist accommodation in Samoa is being threatened by accelerated coastal erosion. 76
4.6 Artificial (i.e. man-made) hotel beach at Coconut Bay, Tobago; the beach is approximately 200 m in length and 60 m in width,

	and was constructed using limestone sand imported from Guyana. Note also the groyne, constructed in an attempt to slow down the rate of sand loss. The developer of the hotel, John Jefferis, thought of the concept and subsequently designed and constructed the beach.	76
4.7	The Giant Panda in China is a symbol of global nature conservation (e.g. the Worldwide Fund for Nature), and climate change is seen as one factor that puts the species at risk.	77
4.8	Rwenzori Mountains National Park, Uganda. Snow cover on the mountain is already reduced as a result of global warming.	79
4.9	Trail map – the trail network and associated infrastructure are being modified as a result of changes in the climate.	79
4.10	Hotel pool on Denarau Island, Fiji.	86
4.11	Wanaka on a sunny winter day.	88
4.12	A recent trend in the Caribbean is to use white rock when constructing groynes and other forms of coastal protection, in order to improve the aesthetics while also endeavouring to ensure the beach has adequate sand cover.	96
4.13	Russian icebreaker *Kapitan Khlebnikov* – retrofitted for passengers in Pond Inlet, Nunavut, Canada, in August 2010.	100
5.1	Eurostar in Brussels.	110
5.2	One of about 600 Safa Tempos in Kathmandu.	115
5.3	Passengers at Gatwick Airport, London. It is the UK's second largest airport and the busiest single-runway airport in the world. It serves over 200 destinations for around 33 million passengers a year.	121
5.4	Solar lamp, lighting the garden environments in a hotel in Tobago.	127
5.5	Greenpeace at the 2007 UNFCCC Climate Summit in Bali, Indonesia.	131
6.1	Mountain landscape around Queenstown and Glenorchy, often used in Tourism New Zealand imaging and branding.	142
6.2	Biking in Kangaroo Valley, Australia.	146
6.3	City bikes for tourists in Oslo, Norway.	148
6.4	One of four double-decker coaches which commenced operations in 2007. It meets Euro 3 Emission Standards. The bus is the largest in New Zealand, and is estimated to carry 23,725 passengers per year – the equivalent of removing 47,000 private vehicle journeys from the roads each year.	156
6.5	Cycle tourism on small country roads in South Germany.	158
6.6	Solar panels on an Ecoventura cruise vessel in Ecuador.	160
6.7	Fahrtziel Natur advertisement by the Deutsche Bahn.	162
6.8	Solar hot water on tourist accommodation in Suva, Fiji.	164
6.9	The Hytte in Northumberland, self-catering accommodation (with turf roof).	165

6.10 Whistler Blackcomb gondola. 167
6.11 Carbon offsetting project by Atmosfair: Cow dung is fermented in underground biogas digesters, producing an almost odourless and high-energy content gas. 176
7.1 Promotion of 'low carbon health promotion' accommodation at Bangkok airport. 181
7.2 Overwater bungalows are an important part of the brand of tourism in the Maldives. 196
8.1 Wind turbine installed at the Jiminy Peak Mountain Resort. 206
8.2 Intercontinental Resort and Thalasso Spa in Bora Bora, French Polynesia. The pipeline carrying cold ocean water from 900 m depth runs across the reef flat (foreground) to a building located just above the shoreline. As is evident, the installation has no significant visual impact. 208
8.3 Beach of La Brée-les-Bains, Île d'Oléron, France. Note how the erosion control works limit the useable area of beach. 221
8.4 View from tourist accommodation at Hermanus. 227
8.5 Aviation fuel pump in a regional airport in Germany. 233

Figures

1.1 International tourist arrivals (in millions) (*2010 are preliminary results). 2
1.2 Focus of the book: policy and practice. 4
1.3 Structure of the book: adaptation responses. 6
1.4 Structure of the book: mitigation responses. 6
2.1 Observed changes in components of the global climate system. © American Meteorological Society. Reprinted with permission. 9
2.2 Observed changes in snow cover and glaciation. © American Meteorological Society. Reprinted with permission. 10
2.3 Annual global average temperature anomalies (relative to 1961–1990) from 1850 to 2010 from the Hadley Centre/CRU (HadCRUT3) (black line and grey area, representing mean and 95% uncertainty range), the NOAA National Climatic Data Center (light grey); and the NASA Goddard Institute for Space Studies (dark grey). 11
2.4 Six-step approach to assessing the climate change impacts on tourism and identifying adaptation options. 24
3.1 The adaptation process, highlighting the importance of a strong enabling environment for effective adaptation in the tourism sector. 32
3.2 Structure and content of Saint Lucia's National Climate Change Adaptation Policy and Strategy. 36

3.3 The approach used in the ARICC Project to identify priority adaptation strategies and measures to increase the weather and climate resilience of rail infrastructure and operations. 51
4.1 Forced and planned adaptation – the two categories of actions designed to reduce the consequences of climate change. Also shown for completeness is mitigation, the third response to climate change. 56
4.2 Adaptation initiatives northern European tourists and Mediterranean tourism service providers might take as a result of an increase in the number of unacceptably hot summer days at Mediterranean destinations. 61
4.3 Weather and related phenomena that can be influential for given phases of flight operations. 69
4.4 Causes of weather-related delays at Heathrow Airport, for the period January, 2001 to May, 2006. 70
4.5 A real time FACET graphical display showing convective weather around Chicago O'Hare International Airport on July 23, 2010. FACET integrates weather and air traffic information, allowing operators to reroute flights to maintain safety and minimize delay. 72
4.6 Winter precipitation trends in the Mediterranean region for the period 1902–2010, shown as departures from the long-term average. 73
4.7 Matrix of approaches to responsible marketing of tourism services in the context of climate risks and their management. 84
4.8 Impact of Hurricanes Hugo and Marilyn on room capacity, visitor arrivals and tourist expenditure in the United States Virgin Islands, 1988 to 1997. 90
4.9 Example of tourism advisory related to a hurricane approaching the Florida Keys. 92
4.10 Queensland Tourism Crisis Management Plan. 93
5.1 Risk-based framework for developing GHG mitigation policies. 106
5.2 Accor Hotels Environment Charter. 107
5.3 Six Senses Virtuous Circle. 109
5.4 Corporate Plan 2009–2011 of the Sunshine Coast Council (Australia). 113
5.5 Energy policies of the European Union. 126
5.6 IATA Conceptual Emission Reduction Scenario. 135
5.7 The four pillars to climate change mitigation by IATA. 135
6.1 (a and b) Cycle paths and tourist attractions in Växjö, Sweden. 144
7.1 The role of monitoring and evaluation in the context of mitigation and adaptation, and of mitigation as a source of financial resources to support adaptation. 181
7.2 The carbon intensity of travel options – a webcomic. Reprinted with permission from Ecometrica. 183
7.3 The three scopes of the Greenhouse Gas Protocol in the context of tourism. 184

7.4 Diagram of the supply chains of Emirates Wolgan Valley Resort in Australia, showing sources of emissions, and highlighting those units that are included in the GHG inventory. 189
7.5 The two main categories of indicators for monitoring and evaluating adaptation to climate change, with examples based on Snowbowl Ski Resort in Arizona, U.S. 194
7.6 Overview of the activities and intended outcomes and impacts of a regional monitoring and evaluation (M&E) system for disaster risk management (DRM) climate change adaptation Caribbean Disaster Emergency Management Agency (CDEMA) and Coordination Unit (CU) in the Caribbean tourism sector. 195
7.7 Illustration of the contrast between maladaptation and actions to protect and restore wetlands to facilitate adaptation to climate change. 198
8.1 Synergies and trade-offs arise when tourism responds to climate change in an integrated manner. 204
8.2 Responses to climate change, from climate change-focused (left) to development-focused (right), with illustrative examples for adaptation, mitigation and the two combined. 205
8.3 Diagrammatic representation of the deep ocean water-cooling system at the Intercontinental Resort and Thalasso Spa in Bora Bora, French Polynesia. (Richard Bailey, Pacific Beachcomber SC Company, Tahiti, French Polynesia) 209
8.4 Pacific Resort Rarotonga, Cook Islands. 214
8.5 Excerpt from the Lusaka Declaration on Sustainable Tourism Development, Climate Change and Peace. 216
8.6 Media coverage of cyclone Wilma (2005) in the Christchurch *Press*, New Zealand. The coverage focuses on negative aspects and safety concerns for tourists. 225
8.7 Hierarchical framework for oil price increase impacts on tourism. 231

Tables

1.1 Domestic and international tourist volumes (2005) worldwide for same-day and overnight visitors. 2
1.2 Natural disasters in 2010 compared with the average of 2000–2009: floods, mass movements (wet) and storms. 3
2.1 Climate change impacts, policy gaps and planned policies for tourism. 14
2.2 Contribution of tourism to global greenhouse gas emissions. 15
2.3 Energy consumption for a wide range of transport modes. 17
2.4 Range of adaptation measures appropriate for ski fields. 21
3.1 New Zealand Tourism and Climate Change Plan: priority areas and actions related to strengthening the enabling environment. 34

3.2 Saint Lucia's National Strategy for Adaptation to Climate Change by the Tourism Sector. 37
3.3 Examples of adaptation actions of relevance to the tourism sector. 44
3.4 Examples of specific adaptation actions and time frames included in the Kakadu National Park Climate Change Strategy 2010–2015. 46
4.1 Changes in climate and their possible consequences for transportation. 65
4.2 Asset and operational risks to Network Rail due to heat, including exacerbation by climate change. 67
4.3 Adaptation measures being used by tourism operators in Wanaka and Queenstown. 88
4.4 Managing tourist satisfaction, health and safety. 94
5.1 Examples of emission reduction targets relevant to tourism. 104
5.2 Descriptions of selected GHGs abatement policy instruments. 118
5.3 Climate change mitigation policies in Australia. 119
5.4 APD rates of duty from 1 November 2009. 122
5.5 Impacts of the suggested energy policies for Barbados, based on detailed energy audits and analysis. 128
6.1 Examples of mitigation measures for tourism. 140
6.2 Overview of mitigation technologies and practices from the transport sector (non-technological measures such as lifestyle changes are not included). 149
6.3 Aviation's contribution to anthropogenic climate change. 150
6.4 Emission reductions with technologies currently available. 152
6.5 Biofuel flights undertaken by various airlines. 154
6.6 Comparison of offset providers from New York to the Caribbean. 175
7.1 National GHG account for tourism, Australia. 186
7.2 Elements of the energy system that determine energy consumption and GHG emissions in Whistler, Canada. 188
7.3 Proposed climate change related indicators for sustainable tourism destinations in the Caribbean. 195
7.4 Extract from the Results Framework for the Project 'Increasing Climate Change Resilience of Maldives through Adaptation in the Tourism Sector'. 197
8.1 Energy Balance Sheet for Jiminy Peak Mountain Resort, Massachusetts, US. 207
8.2 Summary of strategic directions for the beach destination scheme of the Oléron Island Beach Plan. 219

ACRONYMS

AICST	APEC International Centre for Sustainable Tourism
APD	British Air Passenger Departure Duty
APEC	Asia-Pacific Economic Cooperation
CAST	Caribbean Alliance for Sustainable Tourism
CERs	Certified Emission Reductions
CFCs	Chlorofluorocarbons
CSR	Corporate social responsibility
ETS	Emissions Trading Scheme
EU	European Union
GHG	Greenhouse gas
IAPAL	International Air Passenger Adaptation Levy
IATA	International Air Transport Association
ICAO	International Civil Aviation Organisation
ICLEI	International Council for Local Environmental Initiatives
IPCC	Intergovernmental Panel on Climate Change
ICZM	Integrated coastal zone management
KSI	Key sustainability indicator
OTA	Oceanic Tailored Arrivals
RES	Renewable energy supply
UNEP	United Nations Environment Programme
UNWTO	The United Nations World Tourism Organisation
WMO	World Meteorological Organisation
WTTC	World Travel and Tourism Council

1

INTRODUCTION

1.1 Tourism and climate change

There is increasing awareness of the importance of climate change for tourism, but detailed understanding is still limited, as are practical and effective responses. Tourism is now back on its path of growth, while at the same time the adverse consequences of climate change are accelerating. The United Nations World Tourism Organisation (UNWTO) reports that, after a significant drop in international tourism due to the global recession, global arrivals in 2010 exceeded the historic high of 2008. 2010 saw 935 million international tourists worldwide (Figure 1.1). This recovery was much faster than initially expected. The recovery has been led by emerging economies, especially in Asia, with growth rates of 8 per cent relative to 2008, compared with only 5 per cent for advanced economies (e.g. in Europe). Interestingly, tourist expenditure has not recovered at the same level, indicating that those tourists deciding to travel internationally are less willing or able to spend the amounts they did before the global financial crisis. The 2011 events of devastating floods in Australia, the Christchurch (New Zealand) earthquake and the Japanese earthquake, tsunami and nuclear crisis, as well as increasing fuel prices, are likely to have influenced tourist arrivals and flows for 2011, and beyond.

Global tourism statistics typically focus on international tourism, but Table 1.1 below shows that domestic tourism is the main contributor to the size of the sector, globally. Added together, there were about 9.8 billion tourist arrivals worldwide in 2005 (no later comprehensive assessment is available), of which only 1.75 billion were international in nature. The sheer volume of global tourism indicates its relevance for climate change, in terms of both its contribution to greenhouse gas (GHG) emissions and its potential vulnerability to adverse weather and climate impacts. As evident in earlier research (e.g. Becken, 2007, 2008a, 2008b; Gössling *et al.*, 2002), the main concern about tourism's GHG emissions relates to transport. When considering

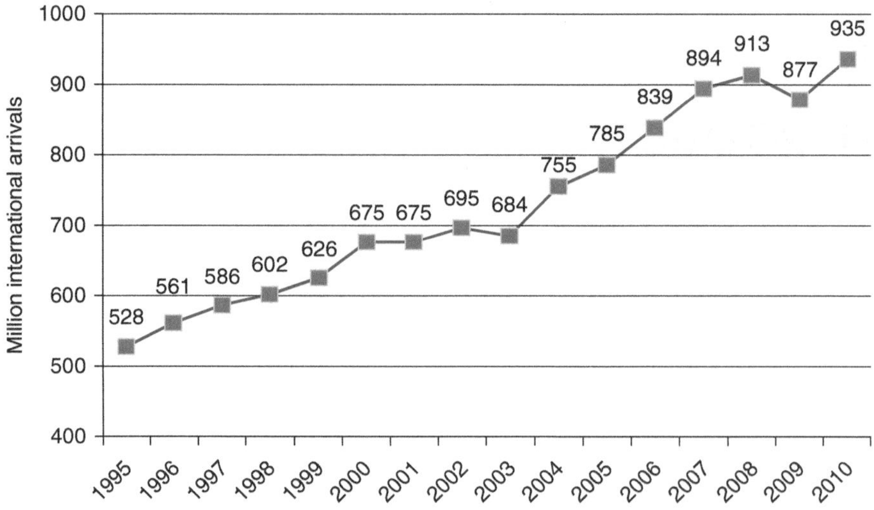

FIGURE 1.1 International tourist arrivals (in millions) (*2010 are preliminary results). (UNWTO, 2011)

TABLE 1.1 Domestic and international tourist volumes (2005) worldwide for same-day and overnight visitors. (Becken, 2009, after UNWTO *et al.*, 2008)

		Arrivals (billions)		
		Domestic	*International*	*Total*
Same-day	Over land/water	4.0	0.99	5.0
	By air	0.04	0.01	0.05
Overnight	Over land/water	3.52	0.41	3.93
	By air	0.48	0.34	0.82
	Total	8.04	1.75	9.8

overnight tourism (some consider same-day visitation to be part of recreation rather than tourism), tourism's proportion of air travel is 17 per cent across all tourists and 45 per cent for international arrivals. The high reliance on air travel by international overnight tourists is one reason why there has been significant focus on this mode of transport in climate change related tourism research (e.g. Peeters *et al.*, 2004; Pentelow and Scott, 2011).

2010 was a year of climate extremes. The National Oceanic and Atmospheric Administration (NOAA, 2011) summarized important observations for the year:

- The global land surface temperatures were the second warmest on record, and 0.96°C above the twentieth-century average.

- 2010 was the wettest year on record.
- While the 2010 Pacific hurricane season had the fewest hurricanes on record since the mid-1960s, the Atlantic season had the second-most hurricanes on record.
- The Arctic reached its third smallest annual sea ice minimum on record, behind 2007 and 2008.
- Record cold and major snowstorms with heavy accumulations occurred across much of eastern North America, Europe and Asia.
- There was an unprecedented two-month heatwave in Russia.
- There were devastating floods in Pakistan.

Of the 373 recorded natural disasters in 2010, 344 (92.2 per cent) were of a climatological or hydro-meteorological nature (Ferris and Petz, 2011). The 182 floods reported in 2010 affected almost 180 million people and killed more than 8,100. This is almost double the average number of people affected every year by floods during the 2000–2009 period (Table 1.2).

An authoritative assessment from the Intergovernmental Panel on Climate Change (IPCC) published in 2011 found that by the end of the twenty-first century there will likely be increases in the frequency of both extreme high temperatures and heavy precipitation events, with the latter contributing to increases in local flooding in some catchments or regions. While maximum wind speeds in tropical cyclones are likely to increase in some ocean basins, the global frequency of tropical cyclones will either decrease or remain essentially unchanged. A pole-ward shift in mid-latitude storm tracks is anticipated. Due to reduced precipitation and/or increased evapotranspiration, droughts will intensify in the twenty-first century in areas such as the Mediterranean, southern North America and southern Africa. Most glaciers will continue to retreat. Extreme high sea levels will occur with

TABLE 1.2 Natural disasters in 2010 compared with the average of 2000–2009: floods, mass movements (wet) and storms. (Ferris and Petz, 2011)

	Floods	*Mass movements (wet)*	*Storms*
2010 recorded	182	29	83
2000–2009 average recorded	173	19	105
2010 killed	8,119	3,258	1,367
2000–2009 average killed	5,401	763	17,222
2010 affected (million people)	178.0	2.5	8.0
2000–2009 average affected (million people)	94.8	0.15	39.7

Note: for 2010 including 25 landslides and four avalanches; for 2000–2009 data including landslides, avalanches, rockfalls and subsidence.

increasing frequency as a result of the combination of increasing mean sea level and the likely increase in tropical cyclone maximum wind speed.

The combination of tourism growing in importance, numerically and economically, and the current and anticipated changes in climate, highlight the urgent and immediate need for strong and effective responses both to reduce tourism's contribution to climate change and to the threats and opportunities climate change represents for the tourism sector.

1.2 Rationale and structure of the book

This book seeks to address the gap between our increasing understanding of the drivers and consequences of climate change on the one hand and our access to very specific practical information, including examples of action (Figure 1.2), on the other. The book therefore positions itself between the rapidly growing volume of literature presenting higher-level assessments of climate change and the implications for tourism (e.g. Becken and Hay, 2007; Gössling, 2010; Gössling and Upham, 2009; Jones and Phillips, 2011), and the guides that are often produced by industry organizations and typically have a very specific focus (e.g. The International Tourism Partnership, 2010). Thus, while explicitly building on our earlier book (Becken and Hay, 2007) and even more recent publications, this present book will focus not only on the roles of supportive policies and institutions in ensuring a strong 'enabling environment' for practical responses, but also on the practical responses themselves.

At the 2010 World Travel Market in London, the World Travel and Tourism Council (WTTC) and the world's leading private-sector tourism businesses presented a framework for true partnerships in the creation of policies that address climate change and enhance the resilience of tourism businesses to climate change (WTTC, 2010). This is an example of the linkages and initiatives that are the focus of this book. More specifically, the book brings together a large number of case studies to illustrate how tourism policy and industry initiatives have been implemented, and considers factors that influence their success. These case studies have been selected to showcase good practice, innovation and leadership, but they are also analysed critically to identify potential shortcomings or lessons learned for future implementation. In addition to the case studies within the text, tables summarize further examples that illustrate practical aspects of policy or implementation. These examples are briefly summarized and a web link is provided for readers who wish to obtain further details.

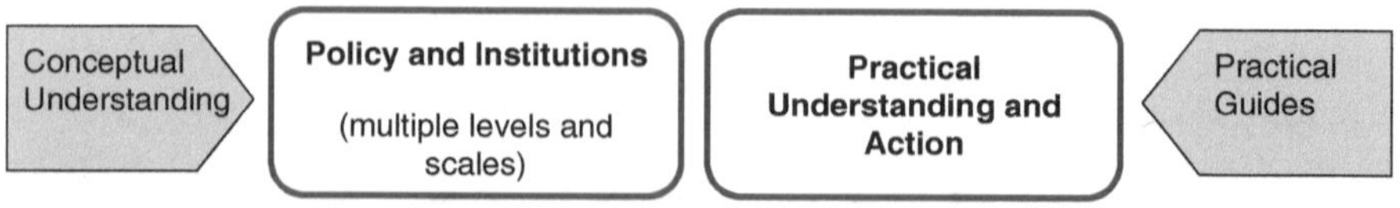

FIGURE 1.2 Focus of the book: policy and practice.

This book is based on the growing realization that both policy and practical responses are required if tourism is to deal with climate change effectively. The two levels of response are not independent of each other. Without such policy, whether formal or informal, practical interventions tend to be random, uncoordinated and ad hoc. Policies are developed by government, private sector organizations, businesses, groups – e.g. non-governmental organizations (NGOs) – and individuals. There are many different views on what policy exactly is, but for the purpose of this book it is assumed that policy in the wider sense relates to:

- a set of principles or rules;
- an overall high-level plan setting out goals and procedures;
- a guide for decision making; and
- a particular approach, selected from a range of alternatives.

Policies developed by government organizations are referred to as 'public policy' and describe the general 'course of action', including specific regulations, incentives and information campaigns.

The book is structured as follows. Following this introductory chapter, a synthesis of the current understanding of climate change and tourism is provided (Chapter 2, Recent Advances in Knowledge and Understanding). The main part of the book focuses on how policy leads to good practice. Initially it deals with adaptation and mitigation separately, first looking at the policy level and then the practical experience. Following the detailed discussions of adaptation (Chapter 3, Adaptation Policy, Planning and Institutions and Chapter 4, Adaptation Practice) and mitigation (Chapter 5, Mitigation Policy and Chapter 6, Mitigation Practice), the book dedicates a chapter to monitoring and learning (Chapter 7, Assessing Performance, and Learning). Chapter 8, Integrated Responses to Climate Change, focuses on cross-cutting aspects such as how to benefit from synergies between mitigation and adaptation, between climate change, tourism and sustainable development, and between disaster risk reduction and climate change adaptation. The relevance of energy security for climate change and tourism is also discussed. Finally, the concluding chapter (Chapter 9, Conclusion) assesses the overall content of the book and explores the future of tourism and climate change from research, policy and operational perspectives.

The more detailed structure of the core of the book is visualized for adaptation (Figure 1.3) and mitigation (Figure 1.4). The adaptation framework begins, conceptually, with an acknowledgement of the relevant climatic factors and anticipated changes that present risks and opportunities to tourism. These require interventions, at both policy and practical levels. Detailed analyses (e.g. of implications for destination appeal and transportation) are organized in five different sub-sections, building on the success of the analytical framework used in our earlier book. Findings from Chapter 7, the Assessing Performance, and Learning chapter, provide the opportunity to improve responses in the future.

The analysis and discussion of mitigation initiatives begins with an assessment of policy responses (Figure 1.4). These are grouped into policies at a local, national,

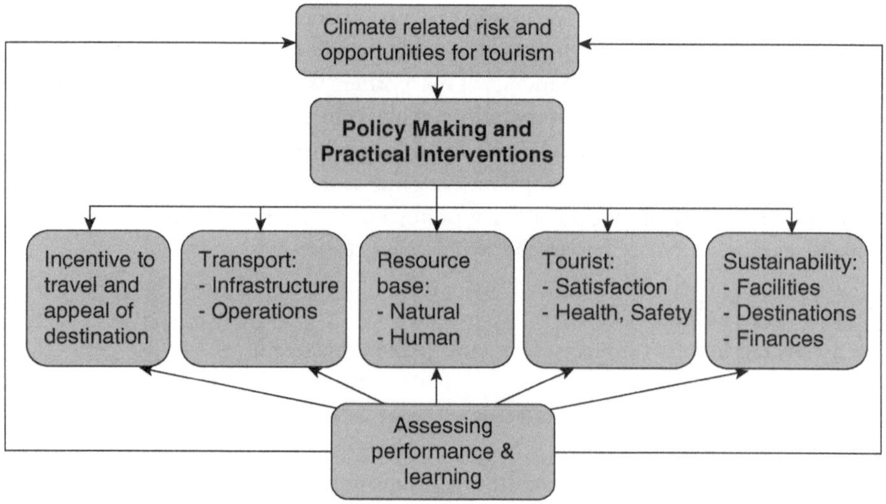

FIGURE 1.3 Structure of the book: adaptation responses.

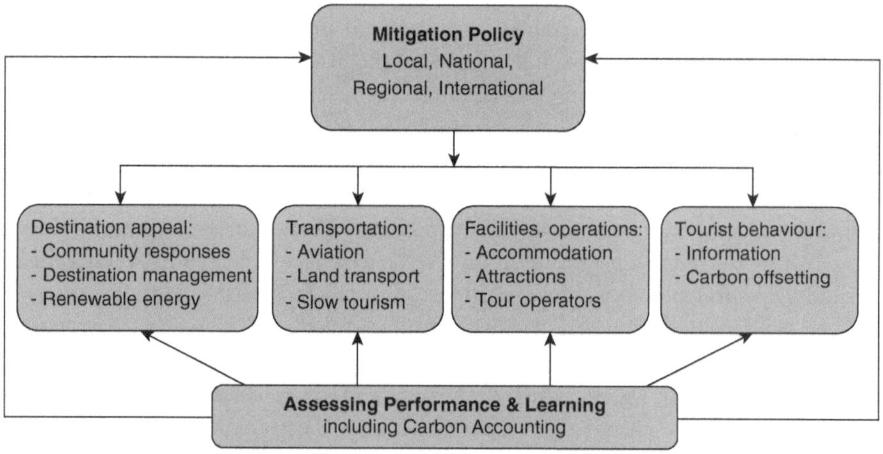

FIGURE 1.4 Structure of the book: mitigation responses.

regional (e.g. European Union) and international level. The chapter on practical mitigation interventions (Chapter 6) begins with initiatives reported at the destination (or community) level, followed by an analysis of initiatives from the transportation sector. The third section in the Mitigation Practice chapter presents examples of climate change mitigation at the business level, including accommodation providers, operators of tourist attractions and tour operators. The last section deals with tourist behaviours and how tourists can contribute to reducing greenhouse gas emissions. Again Chapter 7, Assessing Performance, and Learning (e.g. through carbon accounting), will feed back to the mitigation initiatives undertaken at all levels.

The overall Conclusion (Chapter 9) synthesizes the lessons learned from the book and makes recommendations for future climate change policies and actions in the tourism sector.

2

RECENT ADVANCES IN KNOWLEDGE AND UNDERSTANDING

This chapter provides a synopsis of knowledge and understanding related to the interactions between tourism and climate change, especially that which has been generated over the last five years. The aim of this chapter is to provide up-to-date information on the topic, as well as a conceptual understanding that provides the basis for the more practical chapters that follow.

2.1 Climate change science

Recently reported results of monitoring (e.g. WMO, 2011) and modelling (e.g. Hirabayashi *et al.*, 2008) provide high confidence that significant changes in atmospheric and oceanic conditions are occurring and that these changes will continue well into the future. Changes have been observed not only in the mean conditions (Figures 2.1 and 2.2) but also in both the observed frequency and intensity of weather and climate extremes such as heavy rainfall, frequently resulting in flooding; droughts; high sea levels, often exacerbated by storm surges; and possibly cyclones. These trends in mean and extreme conditions are consistent with those anticipated as a result of global warming (IPCC, 2011).

The decade 2001–2010 was the warmest on record, with average temperatures being 0.46°C above the 1961–1990 mean and 0.21°C warmer than the previous record decade 1991–2000. In turn, 1991–2000 was warmer than previous decades, consistent with the long-term warming trend shown in Figure 2.3 (WMO, 2011).

In its Fourth Assessment Report the Intergovernmental Panel on Climate Change (IPCC, 2007a) identified and assessed the likelihood of future changes in the climate. Continued GHG emissions at or above current rates would very likely induce changes to the global climate system by the end of this century that are larger than those observed over the last century. Just 20 further years of GHG emissions at present-day levels would give a 25 per cent probability that the average

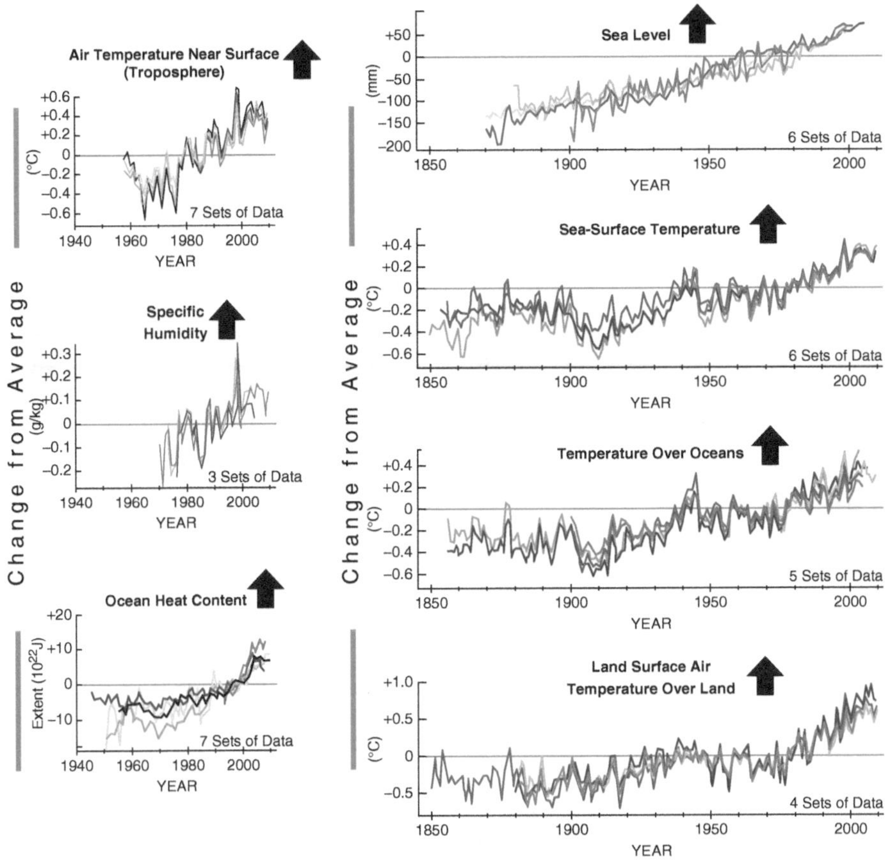

FIGURE 2.1 Observed changes in components of the global climate system. (Arndt *et al.*, 2010) © American Meteorological Society. Reprinted with permission.

temperature of the Earth would increase by more than 2°C. Projections to the end of the twenty-first century include:

- Warming is expected to be greatest over land, and at higher northern latitudes.
- Snow cover and sea ice are expected to shrink in both polar regions.
- It is very likely that hot extremes, heatwaves and heavy rain events will become more frequent.
- It is likely that tropical cyclones/hurricanes/typhoons will become more intense.
- Storm tracks outside of the tropics are likely to shift pole-ward, resulting in changes to wind, rain and temperature patterns.

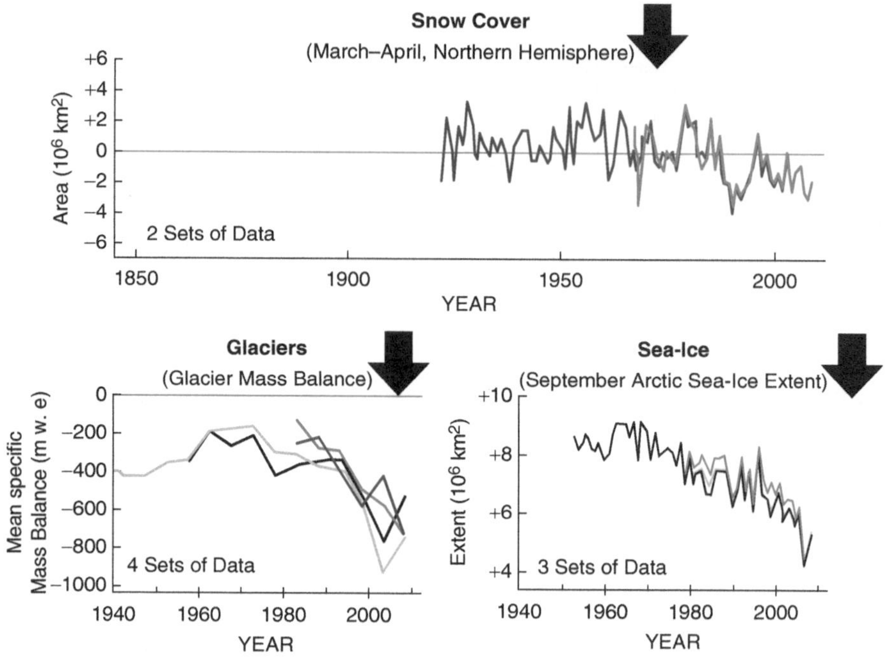

FIGURE 2.2 Observed changes in snow cover and glaciation. (Arndt *et al.*, 2010) © American Meteorological Society. Reprinted with permission.

- There are very likely increases in precipitation at high latitudes.
- There are likely decreases in precipitation at subtropical latitude land areas.

Since that report, several studies (e.g. Allison *et al.*, 2009; Richardson *et al.*, 2011; Rummukainen *et al.*, 2010; Climate Commission, 2011; IPCC, 2011) have shown that the IPCC Fourth Assessment (IPCC, 2007a) findings were in fact conservative, and that many components of the climate are already changing near or even beyond the upper ranges of the IPCC projections. For example:

- Globally there has been an overall decrease in the number of cold days and nights, and an overall increase in the number of warm days and nights.
- Surface melting of the Greenland Ice Sheet is accelerating, permafrost is thawing more quickly than previously thought, and Antarctic ice shelves are collapsing more rapidly than expected.
- Summer-time melting of Arctic sea ice has accelerated and is now exceeding even the worst-case scenarios, with the area of sea ice melt during 2007–2009 being about 40 per cent greater than the average IPCC projection.
- Recent observed increases in global average sea level (3.4 mm/yr over the past 15 years) are approximately 80 per cent greater than past IPCC projections.

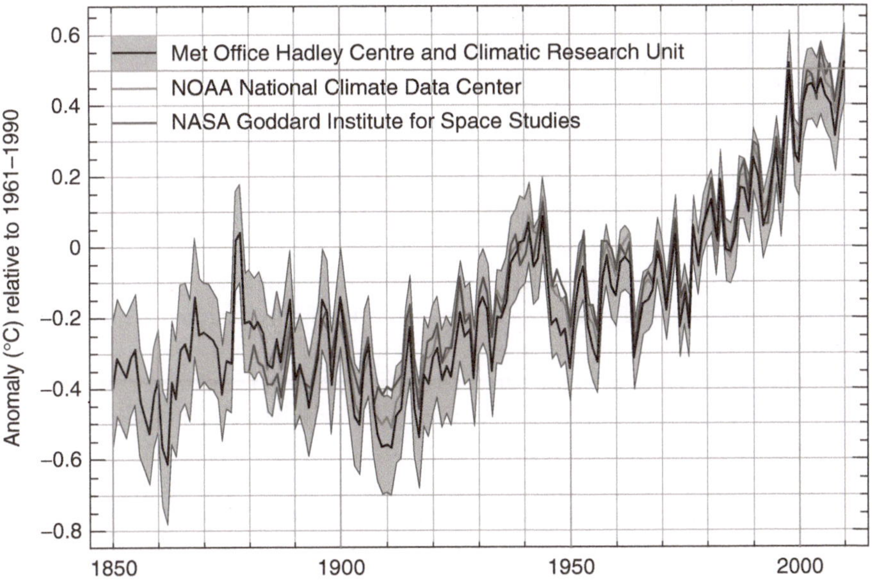

FIGURE 2.3 Annual global average temperature anomalies (relative to 1961–1990) from 1850 to 2010 from the Hadley Centre/CRU (HadCRUT3) (black line and grey area, representing mean and 95% uncertainty range), the NOAA National Climatic Data Center (light grey); and the NASA Goddard Institute for Space Studies (dark grey). (Met Office Hadley Centre, UK, and Climatic Research Unit, University of East Anglia, United Kingdom) (WMO, 2011)

- Recent estimates of the future sea-level rise indicate that by 2100 global sea level is likely to rise by at least twice the IPCC projections, and if future emissions are unmitigated it may well exceed 1 metre; the upper limit has been estimated as approximately 2 metres sea-level rise by 2100.

2.2 The climate change debate

There are two sides to the debate on climate change: on the one hand, sceptics claim that the science is flawed and accuse scientists of 'hiding' data that contradict expected findings, and eagerly leap on any admission of gaps or uncertainty in climate change data and projections of future changes. Often the media finds these alternatives to the mainstream views on climate change more suited to the increasingly 'sensationalist' nature of modern media. On the other side of the debate, scientists devote considerable time and effort to filling gaps and developing more accurate models, but find it difficult to openly communicate any shortcomings for fear of fuelling the sceptics' claims. 'The climate of suspicion' is a phrase used to describe this situation (Schiermeier, 2010).

The divide between the public- and science-dominated views on climate change is often highlighted in opinion polls and surveys of peer-reviewed and popular media reports (Hoffman, 2011). For example, while 84 per cent of scientists surveyed consider the Earth is getting warmer because of human activity such as burning fossil fuels, just 49 per cent of the public held that view (Pew, 2009). A poll conducted in September 2011 found that 20 per cent of Americans, 20 per cent of Britons and 14 per cent of Canadians think 'global warming is a theory that has not yet been proven'. On the other hand, 43 per cent of Britons, 49 per cent of Americans and 52 per cent of Canadians consider that 'global warming is a fact and is mostly caused by emissions from vehicles and industrial facilities' (Angus Reid, 2011).

Certainly, there are gaps in the current knowledge about the climate system, how it may be changing, and which of these changes may be caused or amplified by human activity. In addition, in climate science the most critical information is sometimes the least reliable. For example, sea-level predictions critical to low-lying coastal areas have previously been based on information that many in the scientific community have found to be lacking – although this information is being improved all the time.

There are also limitations within climate change models that mean making relevant projections is highly challenging. Global climate models utilize data that can track, and therefore project, global patterns and changes. They are less useful for making predictions on smaller, country- or community-based scales. The weaknesses of climate change models include a limited ability to simulate aspects of the hydrological cycle, such as estimating vertical air movement, a key element in the transferring of humid air in the tropics to other parts of the planet. There are also uncertainties around the true nature and extent of natural variability. Using global data for more localized projections can in fact exaggerate any inherent weaknesses in the global model. For this reason, most experts are reluctant to make predictions at sub-regional levels. Such uncertainties and limitations make adaptation planning and decision making challenging for local policy makers and planners.

2.3 Policy for climate change and tourism

Tourism stakeholders are now becoming more involved in planning for, and responding to, climate change. Reasons for the limited engagement in the past include low awareness of the relevance of climate change to the tourism sector and how the consequences might be best addressed, the limited capacity of public sector bodies, and lack of coordination between relevant agencies (Becken and Clapcott, 2011; OECD and UNEP, 2011). The tourism sector tends to focus more on marketing and the facilitation of economic growth. Such ambitions for growth are often seen as being at odds with national or international emission reduction goals. National climate strategies, such as that prepared by Sweden, do not pay attention to tourism and its growth. Such strategies may therefore fail to meet their goals (Gössling and Hall, 2008). An analysis of tourism-specific responses to climate change in Australia found that they are more evident in those states where strong climate change policy frameworks are prevalent (Zeppel and Beaumont, 2011).

Frequently the need to reduce the impacts of climate change on tourism is considered an issue for the future. This ignores the fact that many tourism investments and decisions made today will have an impact for decades, by which time entirely different and riskier climate conditions might well prevail (UNWTO *et al.*, 2008). Specific challenges and opportunities for engaging tourism stakeholders in climate change adaptation were highlighted by Nicholls and Holecek (2008) in a study on tourism in Michigan, United States, and by Turton *et al.*, 2010 for four Australian case studies. Developing countries are also starting to become involved in policy making and planning for climate change and tourism, as evidenced in the active participation of government officials from several African countries in the 2011 International Institute of Peace through Tourism conference in Zambia. This meeting focused on addressing climate change at both political and practical levels.

A recent analysis of policy making for climate change by tourism organizations in Fiji and New Zealand highlighted that many of the existing (non-tourism) climate change policies are conceptually adequate, but need to be enhanced by incorporating tourism-specific considerations (Becken and Clapcott, 2011). The study also noted that a more rigorous enforcement of existing policies, such as building standards or set backs for beach-front developments, would contribute to preparing the tourism sector for climate change. A lack of enforcement of existing policies has also been identified as a key barrier to addressing climate change in other parts of the world, for example in the Greater Accra Region (Ghana) for hotel energy efficiency (Mensah, 2006) and for both mitigation and adaptation in island states in the Caribbean (Yaw, 2005). Other important barriers faced by tourism stakeholders in Canada relate to a lack of publicly available and easily understandable information, insufficient coordination between government departments, and poor coordination between government decision makers, marketing organizations and the tourism industry (Dodds and Graci, 2009). Avoiding unsustainable development, such as the construction of inappropriate, low-resilience tourism buildings in vulnerable locations, is often the first step in adapting to climate change. Table 2.1 provides examples of the types of policies and actions that tourism policy makers could consider to address climate change impacts and policy gaps.

Recent research on climate change policy and tourism has attempted to assess the impacts of global climate policies on tourism demand (e.g. Mayor and Tol, 2007, 2008, 2010). Gössling *et al.* (2008) examined the impact of different aviation policies on airfares and arrivals at small island destinations. As confirmed by Pentelow and Scott (2011), the study found that only very high costs of carbon would result in significant impacts on tourist arrivals. In their more specific study of the Caribbean, Pentelow and Scott (2011) also compared the impacts of the proposed International Air Passenger Adaptation Levy (IAPAL) and concluded that, while the impacts on arrivals are similar, the IAPAL is preferable to a carbon tax as it provides funds for adaptation for the visited destinations (in the case of developing country destinations).

TABLE 2.1 Climate change impacts, policy gaps and planned policies for tourism. (After Becken and Clapcott, 2011).

	Decrease in tourist arrivals	*Poor performance by businesses*	*International aviation*	*Research gaps*
Key impacts	• Lower or negative growth rates • Undermining of the national brand	• Exposure of bad practice (reputational impacts) • Carbon inefficiency undermines yield • Lack of innovation • Lack of future oriented planning and investment	• Technological improvements are being made but are not sufficient to outweigh growth impacts • Media coverage on 'air miles' may reduce travel • Policy measures increase cost of travel	• Emergence of new markets • Anticipated climate impacts on tourism • Changing tourism yield
Policy gaps	Lack of understanding of tourist perceptions and how these translate into actions. Lack of integration with other export-related policies.	Tourism-specific information on energy efficiency is limited. Voluntary measures are not wide-spread. Initiatives are fragmented and disconnected.	International negotiations on emissions from aviation have been unsuccessful so far. Tourism has not been linked well to aviation initiatives. Some implement their own national policies.	Lack of future oriented tourism research. Little understanding of climate change impacts on destinations. Insufficient innovation to cope with future changes.
Possible policies and actions	Monitoring system to track tourist perceptions; compare with actual behavioural changes. Establish a process for disseminating market intelligence to the sector. Monitor offshore policy environment and work closely with other government agencies at an international policy level.	Provide practical guidance and information on energy efficiency and relevant policies to tourism businesses. Analyse if regulatory framework is sufficient. Provide tools for carbon footprinting and assist businesses in risk management. Develop a carbon offsetting standard. Integrate industry and public sector initiatives (e.g. local government).	Encourage international partnerships that help reduce aviation emissions. Encourage an integrated approach across the aviation sector industry and agencies to generate initiatives that reduce emissions across the whole journey. Enhance information exchange between relevant agencies, e.g. Ministry of Tourism and the Ministry of Transport.	Establish a research strategy and funding stream for tourism and climate research. Integrate tourism research with that is already occurring in other areas. Facilitate communication between researchers and end-users. Use scenario planning to chart climate change futures, indicators and responses.

2.4 Measuring greenhouse gas emissions by tourism

The last five years have witnessed a number of studies that attempted to measure the GHG emissions from tourism. The common research approaches include energy audits and 'carbon footprinting' (i.e. accounting for emissions of carbon dioxide and other greenhouse gases), ecological footprinting, and life cycle analysis (Filimonau *et al.*, 2011). A global analysis was undertaken by an expert team on behalf of the UNWTO, and published as a technical state-of-the art report for the Davos Climate Change and Tourism Conference in 2007 (UNWTO *et al.*, 2008). The analysis comprised an emissions estimate for tourism in 2005 and projections into the future (under business-as-usual and different scenarios) for 2035 (Table 2.2). It was found that the 1,307 million tonnes of carbon dioxide (CO_2) emitted by tourism in 2005 represented about 5 per cent of the world's CO_2 emissions. There are no estimates on tourism's emissions that are more recent, but these are likely to have grown annually, except during 2009 when the global recession resulted in a decrease in travel demand.

The key contributor to tourism emissions is clearly transportation. Globally, 75 per cent of tourism's emissions are due to transport (Table 2.2). In many countries tourism is a significant contributor to national transport emissions. For example, tourism transport makes up 12.3 per cent of New Zealand's transport emissions (Becken, 2008b). However, despite such significant contributions, in national transport strategies tourism is often overlooked as a potential target market for climate change mitigation initiatives.

The energy and carbon intensity of different transport modes has been assessed and compared in many studies. Some studies consider the climate change impact of transportation in comparison with other external costs (e.g. Maibach and Schneider, 2002; Peeters *et al.*, 2007), in relation to the transport mode's life cycle, and wider supply chains (Chester and Horvath, 2009). Transport energy efficiencies are often reported in different units (e.g. miles per gallon, or kilometres per litre, or litres per kilometre) and assessments also use different frames of reference. A common way of measuring efficiency, for example, is energy-use per seat-kilometre. This measure

TABLE 2.2 Contribution of tourism to global greenhouse gas emissions. (UNWTO *et al.*, 2008)

	2005		*2035*	
Sub-sectors	*CO_2 (Mt)*	*%*	*CO_2 (Mt)*	*%*
Air transport	515	40	1,631	53
Car transport	420	32	456	15
Other transport	45	3	37	1
Accommodation	274	21	739	24
Activities	48	4	195	6
Total	1,307	100	3,059	100

divides the total energy consumed by a vehicle to travel one kilometre by the number of seats available (e.g. 400 on a large aircraft). Energy-use per passenger-kilometre, in contrast, takes into account average load factors. The latter will typically result in a higher number than for energy-use per seat-kilometre – unless, of course, occupancy is 100 per cent. As a result, the different ways of communicating transport or vehicle efficiency may be confusing at times, even to the expert. Therefore caution needs to be exercised where results are being presented in a way that could paint the most favourable impressions of the transport provider. As Table 2.3 shows, train travel is typically the most energy efficient transport mode, while air travel and cruise ships are less efficient.

Several countries and destinations have been assessed to determine their 'carbon footprint'. Examples include an analysis of the 'Dutch Holiday Carbon Footprint', which provides detailed information on GHG emissions from Dutch tourists in 2009. The research revealed that emissions increased by 16.5 per cent between 2002 and 2009, largely due to an increased use of the aeroplane for intercontinental holidays from the Netherlands (de Bruijn *et al.*, 2011). Other examples of national GHG assessments for tourism include those prepared for Australia (Forsythe *et al.*, 2008) and Switzerland (Perch-Nielsen *et al.*, 2010). While much attention is paid to international tourism, Becken's (2008b) study in New Zealand focused on the GHG emissions from domestic tourism. Domestic tourism, which is much larger than international tourism in terms of visitor nights in New Zealand, and in many other countries as well, was estimated to emit 1.9 million tonnes of CO_2-equivalent annually. Of this, 60 per cent of the emissions were attributed to overnight tourist transport, 36 per cent were from day-trip transport, and 4 per cent were due to tourist accommodation. A comparison with GHG emissions from international tourists in New Zealand showed that domestic tourism contributes 75 per cent of all tourism emissions within the national borders (i.e. excluding international aviation). On a per-day basis, domestic tourism is more carbon intensive than international tourism. This is typically explained by the shorter length of stay and the relatively higher transport component as a result.

The last few years have seen an increase in research activity into energy use and emissions related to tourism in China. It has been estimated that tourism in China (with over 1,900 million domestic tourists) consumed about 428.30 PJ of energy in 2008, or about 0.51 per cent of the total Chinese energy use. Resulting emissions were estimated to be 51.34 Mt of CO_2 (about 0.86 per cent of the total) (Wu and Shi, 2011). Specific aspects, such as the increase of electricity consumption due to growth in casino tourism, have also been investigated, for example for the case of Macao (Lai *et al.*, 2011).

A destination-based footprint analysis was undertaken for the tourist resort of Davos in Switzerland (Walz *et al.*, 2008). The results show that the annual CO_2 emissions per capita in Davos are 25 per cent above the Swiss average of 6 tonnes per capita, partly because of the energy-intensive hotel sector. The main source of emissions is heating in winter. It was suggested that building renovations and better use of renewable energy sources could reduce the footprint substantially. Travel to the

TABLE 2.3 Energy consumption for a wide range of transport modes.

Efficiency	*Data*	*Source*
French Corail train	0.05 MJ/seat-km	Peeters (2010)
TGV (French high-speed train)	0.13–0.16 MJ/seat-km	Peeters (2010)
Typical electric car	0.17 MJ/seat-km	Peeters (2010)
Honda Civic Diesel 2.2i-CTDi	0.5 MJ/seat-km	Peeters (2010)
Aviation	1.0–1.5 MJ/seat-km	Peeters *et al.* (2005)
East Japan Railway	0.35 MJ/passenger-km (2004)	www.jreast.co.jp/e/environment/pdf_2005/report2005e_22_23.pdf
Swedish Railways (international travel)	0.74 MJ/passenger-km	Becken and Hay (2007) based on Jorgensen and Sorenson (1997)
Efficient hybrid car	1.08 MJ/passenger-km	http://en.wikipedia.org/wiki/Fuel_efficiency_in_transportation
Commuter rail (U.S.)	1.96 MJ/passenger-km	Transportation Energy Data Book http://cta.ornl.gov/data/index.shtml
Lufthansa fleet (Germany)	1.60* MJ/passenger-km	http://verantwortung.lufthansa.com/fileadmin/downloads/en/LH-sustainability-report-2010.pdf
Intercity rail (U.S.)	1.73 MJ/passenger-km	Transportation Energy Data Book http://cta.ornl.gov/data/index.shtml
Aviation world passenger fleet	1.75* MJ/passenger-km	Macintosh and Wallace, 2009 based on figures provided by Airbus
Camper vans (New Zealand)	2.39 MJ/passenger-km	Becken and Hay (2007)
Domestic air travel (New Zealand)	2.54 MJ/passenger-km	Becken and Hay (2007)
Transit buses (U.S.)	2.78 MJ/passenger-km	Transportation Energy Data Book http://cta.ornl.gov/data/index.shtml
Cruise ship, RMS *Queen Elizabeth 2*	6.08* MJ/passenger-km	http://en.wikipedia.org/wiki/Fuel_efficiency_in_transportation

* Data provided as energy use per passenger per 100 km travelled, converted using a factor of 37.3 MJ per litre for aviation fuel and 36 MJ per litre of diesel.
Note: a (non-peer reviewed) Wikipedia article on fuel efficiencies in transport provides estimates for the energy consumption associated with walking and biking (caloric requirements due to energy burn); these are 0.8 and 0.4 L/100 km, respectively, indicating that cycling is more energy efficient (assuming a moderate speed). (http://en.wikipedia.org/wiki/Fuel_efficiency_in_transportation)

region was not taken into account. Another destination-based study focused on visits to Taiwan's national parks and the related impact on transportation emissions (Lin, 2010). Energy use within a protected area was the focus of an investigation by Sanjay Nepal (2008), who collected data on energy sources and energy-saving technologies from 489 tourist lodges in Annapurna (Nepal). The findings revealed that the primary energy sources of wood and kerosene are increasingly being supplemented by renewable energy sources such as solar hot water and micro-hydro. Other destination-based analyses have employed the method of ecological footprinting. Examples include the contribution of tourist car usage in Lanzarote Island (Martín-Cejas and Sánchez, 2010), tourist itineraries in Shangri-La (China) (Peng and Guihua, 2007), and tourism (including international air travel) in the Seychelles (Gössling *et al.*, 2002). A summary of carbon footprinting analyses for tourism is provided by Sisman and Associates (2007).

The carbon intensity of different types of tourism, or market segments, is receiving increasing attention. For example, in their paper on polar bear viewing tourism in Canada, Dawson *et al.* (2010) estimated that this particular industry contributes 20,892 tonnes of CO_2 per season, considering typical travel distances, length of stay and types of activities (e.g. two tundra vehicle trips). This is about the equivalent of the emissions of 4,000 average (Western) households for a whole year. Eijgelaar *et al.* (2010) examined the GHG emissions of cruise ship tourism to Antarctica. They provide the most accurate estimate of GHG emissions related to cruise tourism worldwide at an estimated 1.5 per cent of global tourism CO_2 emissions. This study also revealed the particularly high levels of GHG emissions related to polar cruise tourism – up to eight times higher per capita and per day than average international tourism trips. Recently, interest has also turned to the climate change impacts of mega-events, for example the Soccer World Cup in South Africa in 2010 (Otto and Heath, 2009). A forthcoming study on space tourism (to be published in *Geophysical Research Letters*) modelled the atmospheric impact of 1,000 suborbital flights departing annually and using a hybrid-powered technology currently being developed by Virgin. These flights would release about 600 metric tons of soot into the stratosphere each year (Matson, 2010). The emissions would have significant impacts, including decreasing mid-latitude temperatures by 0.4°C (due to a 'sunshade' effect) and an increase in warming over both poles by about one degree in the local winter.

2.5 Supply-side responses to climate change

Naturally, considerable research interest has focused on airlines, including how they respond to the issue of climate change (e.g. Bows, 2010; Bows *et al.*, 2009; Morrell, 2009; Sgouridis *et al.*, 2010). Websites such as GreenAir (www.greenaironline.com/Airlines) provide useful industry portals for summaries of the latest research. Such industry information outlets complement the academic literature on aviation and the environment, for example on biodiesel (Wardle, 2003), aircraft energy efficiency (Lee, 2010) and aviation policies (e.g. Hihara, 2010). An ongoing debate relates to the effect of aircraft emissions other than CO_2 on the atmosphere. A recent commentary in *Nature Climate Change* warns that contrails and induced cirrus clouds can cause a greater contribution to anthropogenic warming than previously expected. This paper also provides some ideas for effective mitigation, for example by adjusting flight routes to avoid parts of the atmosphere that are supersaturated with respect to ice (Boucher, 2011).

Recently, research on airline operations and climate change has also started to address wider issues, such as corporate social responsibility (Cowper-Smith and de Grosbois, 2011) and the factors motivating an airline to consider sustainability (Lynes and Andrachuk, 2008). There is some evidence that airline corporate social responsibility is positively linked to an increase in the company's value (Lee and Park, 2010). Another important aspect of aviation and climate change is the relative contribution of aviation to all global emissions, and ethical questions associated with the anticipated growth of aircraft emissions which places a substantial burden on all other sectors (and households) to reduce emissions to 'avoid dangerous climate change' (see Bows, 2010).

Research on other tourism transport modes is relatively scant, except for some recent publications on 'slow travel' (e.g. Dickinson and Lumsdon, 2010). The industry-driven organization, Forum for the Future, is a good example of a holistic guide to sustainable tourism development, highlighting good practice for both climate change mitigation and adaptation, including important aspects of infrastructure development, local communities and financing (Draper and Murray, 2008). Useful suggestions for climate change mitigation from an industry perspective are also provided in a comprehensive report on low carbon tourism prepared by the World Economic Forum (2009).

> For policy makers, it's important to recognize that various Travel and Tourism clusters are at different stages of organizational development, and hence require different policy approaches to accelerate their decarbonization. For example, the global air transport cluster is highly organized, regulated by international aviation rules, and quite advanced in monitoring emissions around the world. On the other hand, the accommodation cluster is fragmented (more than 80 per cent of capacity is owned by independent small to medium-sized enterprises), has few regulations and those that exist differ greatly from one nation to the next, and lacks a system and agreed metrics to monitor global emissions. (World Economic Forum, 2009: 6)

Some detailed studies on climate change and the tourism industry are available for the accommodation sector. Dalton *et al.* (2007), for example, investigated the attitude of accommodation operators to renewable energy supply (RES) in Queensland, Australia. Their results showed that 9.2 per cent of businesses had some form of RES, and that the majority of operators were interested in the potential of renewable energy sources. Larger businesses were more likely to be attracted to RES than were smaller ones. In New Zealand, Becken (in press) undertook an analysis of 16 detailed energy audits of a range of accommodation providers. A key finding was that the average energy savings potential is about 20 per cent, with most of the identified measures having pay-back times of less than one year. Charara *et al.* (2011)

examined water use efficiency for the Barbados hotel sector, an issue that will be increasingly important for climate change adaptation in areas where reductions in precipitation are forecast. The authors concluded that the failure to date to improve water efficiency relates to the fact that water represents less than 5 per cent of annual expenses. A study on environmental innovation in North American ski resorts and hotels provided evidence that the implementation of environmental measures is linked to the perceived simplicity of a measure and its relative advantage (Smerecnik and Andersen, 2011).

With respect to involving tourism businesses and other stakeholders in climate change adaptation, Turton *et al.* (2010) found that the tourism sector is not yet ready to invest in climate change adaptation because of the perceived uncertainties in the magnitude of climate change and related environmental impacts. Often, the ongoing leadership for climate change related issues was seen to rest with the public sector and not with the industry or tourists themselves (Turton *et al.*, 2010), despite the latter being recognized as having the greatest adaptive capacity (UNWTO *et al.*, 2008). Bicknell and McManus (2006) noted that tourism stakeholders involved in the ski industry did not see an urgency to address climate change, as investment cycles for that industry are typically less than ten years. Moreover, adaptation to climate variability (i.e. present-day fluctuations in climate) is often considered sufficient to cope with future climate changes, and therefore concern is limited (for an Austrian example see Wolfsegger *et al.*, 2008). A wide range of adaptation measures (to both variations and changes in climatic conditions) have been identified. These can be usefully classified into measures that: (i) protect the affected business; (ii) expand beyond the affected business; and (iii) share the risks of financial impacts. For an illustrative list of adaptation measures for the ski industry see Table 2.4.

2.6 The environmental behaviour of tourists

Considerable research effort has gone into understanding what tourists want, how environmentally aware or committed they are, and how they value specific environmental attributes. This research is very relevant to both climate change mitigation and adaptation. In their book on 'slow tourism', Dickinson and Lumsdon (2010) provide a highly pertinent review of social and psychological theories (e.g. social dilemma theory) relevant to transport decisions made by tourists. They also present a useful discussion of social practices, habits, identity and status, as well as mobility more widely. Their contribution highlights the difficulties of changing the current behaviours and expectations that tourists attach to travel and mobility. Overall the literature remains ambiguous on the degree to which tourists are aware of climate change and are willing to personally do something about it – for the example of business travellers from Denmark see Lassen (2010).

On the adaptation side, it has been argued that tourists have the greatest adaptive capacity of all those involved in the tourism system (UNWTO *et al.*, 2008). They

TABLE 2.4 Range of adaptation measures appropriate for ski fields. (Hoffmann *et al.*, 2009)

Strategic direction	*Individual adaptation measure*
Protect affected business	Extend artificial snowmaking
	Landscaping and slope development to reduce the required snow cover (e.g. plane slopes)
	Develop higher terrain for slope sports
	Install blankets for snow conservation
	Improve/professionalize slope support
	Concentrate slope sports in higher terrains of the developed area
	Introduce alternative slope activities
	Organize winter sport competitions
	Open slopes with less snow than usual
Expand beyond the affected business (diversify)	Establish/develop hiking trails
	Offer theme hiking
	Organize summer events on the mountain (e.g. open-air concerts)
	Develop bike/downhill routes
	Offer adventure and fun sports (e.g. paragliding)
	Develop gastronomy on the mountain
	Build a summer toboggan run
	Establish/develop snowshoeing trails
	Establish/develop winter hiking trails
	Establish/develop winter toboggan runs
	Organize winter events on the mountain (e.g. concerts)
Share financial risks	Effect a snow insurance
	Split costs (operating costs or investment costs) with local industry
	Split costs with the local authority
	Cooperate with local industry (e.g. joint offerings)
	Cooperate with other ski lift operators (e.g. exchange of machines)
	Merge with other ski lift operators

suggest that three key resources – money, knowledge and time – provide them with relative freedom to avoid destinations impacted by climate change as well as the opportunity to shift the timing of travel to avoid unfavourable weather conditions. Jopp *et al.* (2010) emphasize that this adaptive capacity should not be assumed. Since the final decision whether and where to travel is made by individuals, collectively they are key stakeholders in any adaptation process. Hence it is imperative that their attitudes to proposed adaptation options be well understood, including determining which options are preferred and/or viable, and why. Jopp *et al.* (2010) suggest that such information can be obtained by surveying tourists at destination 'hot spots' and by conducting focus groups.

The existence of 'superficial or veneer environmentalism' (Weaver, 2011: 9) has been noted in a number of studies. McKercher *et al.* (2010), for example, reported that travellers from Hong Kong are generally aware of climate change but show little inclination to voluntarily change their travel behaviour to reduce their carbon footprint. Interest in modifying travel was lowest among those who travelled the most. The authors suggest that changing consumer behaviour may represent the greatest challenge to reducing tourism's GHG emissions in the short to medium term. Another study on polar bear tourism in Canada confirms that tourists are unlikely to engage at a substantial level in climate change mitigation. Dawson *et al.* (2010) found that, despite a general understanding of the impacts of climate change on polar bears, the large majority of tourists did not understand how their own behaviour contributes to, or could help to reduce, future changes in climate.

Market research by Tourism Australia (2009) identified climate change concerns across all major markets, although limited action had been taken by tourists to date. However, the study did point out that a large number of tourists stated that they 'might or will definitely' take climate change into account and encouraged the industry to prepare for it. Such concerns resonate with the findings of Higham and Cohen (2011) in a Norwegian study that indicates long-haul air travel is not necessarily seen as a status symbol any longer, but is increasingly viewed with social disapproval. A social 'tipping point' with regards to the acceptability of long-haul leisure travel might be conceivable in the future.

A number of segmentation analyses have been undertaken to reveal which kinds of tourists might be more amenable to climate change mitigation (Mair, 2011) or which ones might travel with lighter 'environmental footprints' (Dolnicar, 2010: 717; see also Dolnicar *et al.*, 2009). Dolnicar (2010) differentiated between 'environmentally friendly individuals' and 'environmentally friendly tourists', with the former group finding it hard to overcome structural barriers (e.g. different infrastructure) to act as responsibly on holiday as they would at home. This finding confirms Barr *et al.*'s (2010) observation that the holiday situation proves particularly challenging in terms of changing behaviours. A similar situation was observed for tourists to New Zealand, many of whom were unwilling to give up their privilege of travelling. However, these tourists were relatively more inclined to engage in climate change mitigation in their home environment (Becken, 2007).

Segmentation analyses have also been carried out for issues relevant to climate change adaptation. To better understand the values tourists attach to coral reefs, research undertaken in Hawaii, United States, analysed tourist responses to a range of belief statements such as 'humans should manage coral reefs so that humans benefit'. Most respondents fell into a strong protectionist group, particularly for females, snorkellers and swimmers. Scuba divers and anglers tended towards mixed protection–use values (Needham, 2010). Similar research into diving tourism in the Red Sea, Egypt, found that there is a disconnect between the attitudes of dive tourists and tourism suppliers. The latter were less concerned about climate change and may 'lag behind' the environmental awareness of their clients (Marshall *et al.*, 2011). Such information on the beliefs of different user groups can inform the selection and implementation of adaptation measures to protect coral reefs and other tourism resources.

How tourists engage with information they are provided with is another important component of adaptation research. One example is the issuing of weather warnings to tourists as preparation for extreme events so they might protect themselves whilst on holiday (Jeuring, 2011). Another example relates to information on environmental initiatives, for example energy saving, provided by ecotourism operators to their customers. Research in Australia found that businesses that are certified with Ecotourism Australia frequently put climate change mitigation information on their website, not only to differentiate their product but also to educate tourists and increase awareness (Nelson, 2010).

2.7 Tourist perceptions of climate and weather

Changes in climatic conditions are believed to influence tourist behaviour. Accordingly, the body of literature on climate as a resource for tourism has continued to grow, providing more details on which weather and climatic elements are important (e.g. visibility and precipitation, Yu *et al.*, 2009), how destinations, tourists and tourism businesses are affected by climate variability and change (Becken *et al.*, 2010a, 2010b; Kozak *et al.*, 2008; Moreno *et al.*, 2008), and how tourist flows are related to climatic conditions where they live and at their planned destination (Ehmer and Heymann, 2008; Eugenio-Martin and Campos-Soria, 2009).

The Mediterranean has attracted particular attention, simply because it is the most visited destination in the world and there have been concerns that it might become 'too hot' for tourism in the future (Rutty and Scott, 2010). One approach to estimating how climate change will modify tourism flows uses comfort indices. By combining two climate change scenarios and a tourism climatic index, Amelung *et al.* (2007) report that locations where climatic conditions are ideal for tourism are likely to be found further pole-ward in coming decades. For similar reasons destinations such as the Mediterranean will experience shifts in their peak seasons from summer months to the current shoulder periods, while regions in higher latitudes are likely to experience a lengthening of their summer seasons. Similarly, Hein *et al.* (2009) analysed potential climate change impacts on tourism in Spain, and found that the northern and more mountainous provinces might benefit at the expense of the traditional destinations in the south and on the coast.

But the extensive data requirements for estimating changes in climatic attractiveness usually restrict such studies to smaller regions or particular destinations. For example, Endler and Matzarakis (2010) used a composite index of temperature, humidity, wind speed and cloud cover to map tourism potential and thence to explore the differential responses to climate change between low and high elevations in Germany.

Serquet and Rebetez (2011) used empirical data to explore these relationships still further. They found significant correlations between the number of nights spent in 40 resorts in the Alps of Switzerland and the temperatures occurring at lower elevations, with the alpine resorts nearest to cities being most affected by hot temperatures. The relationships are strongest in June and weaker in August than in July. Serquet and Rebetez attribute this to holidays and the peak of domestic tourist demand occurring from July to mid-August, meaning that long-term planned stays dominate more during these months. This compares to June, when tourists already react on a short-term basis to hot days and spend more nights in hotels in mountain resorts. Thus alpine tourist resorts could benefit from hotter temperatures at lower elevations under future climates. Research on tourism in National Parks in Canada also highlights the impacts of climate change on seasonality (Jones and Scott, 2006) and the particular segment of nature-based tourism (Scott *et al.*, 2007). Prideaux *et al.* (2010) propose a six-step model for analysing climate change impacts on the environment and subsequent impacts on tourist behaviour, via changes in environmental perceptions (Figure 2.4).

The climatic conditions where tourists live are also important. A recent study on British outbound flows identified that mean temperature, heatwaves, frost and sunshine are significant variables that affect outbound tourism over time (Rosselló-Nadal *et al.*, 2011). Further, the use of climate and weather information for tourism planning and operations has received increasing attention (Hartz *et al.*, 2006; Scott and Lemieux, 2009; Becken *et al.*, 2011; Jeuring, 2011). The so-called 'weather and

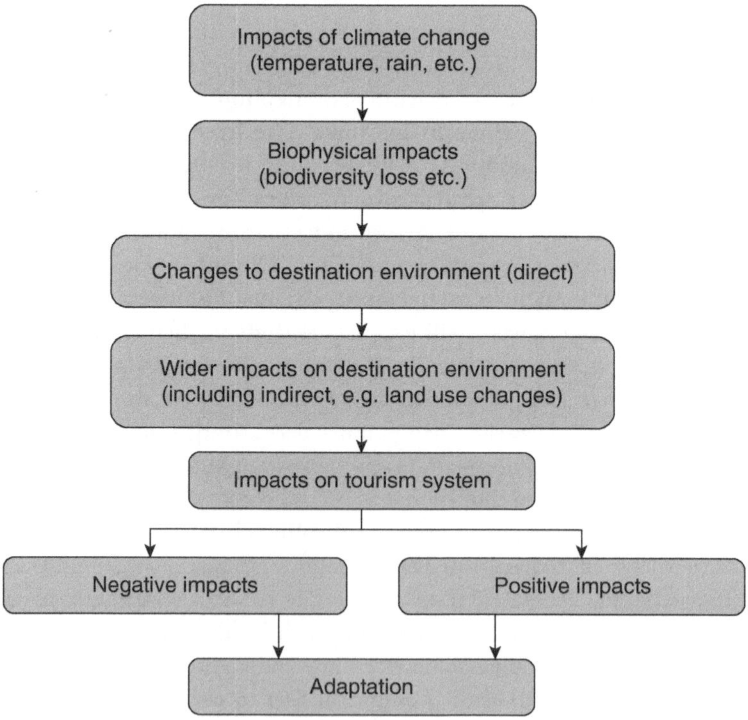

FIGURE 2.4 Six-step approach to assessing the climate change impacts on tourism and identifying adaptation options. (Based on, and simplified from, Prideaux *et al.*, 2010)

climate information system for tourism' has been analysed for the New Zealand situation. Several gaps and shortcomings in the flow of information between suppliers (e.g. the weather information provider), intermediaries (e.g. a regional tourism organization) and end users (e.g. tourism operators, and international and domestic tourists) were identified (Wilson, 2011). Knowledge about climate and weather is an important step towards adaptation, to respond to both climate variability and longer-term climate change.

2.8 Local impacts and destination-level adaptation

Much of the recent research effort on climate change and tourism has dealt with local impacts and adaptation options. This amounted to two thirds of the research output analysed by Weaver (2011). A prominent example of an impact and adaptation assessment is the study undertaken by Turton *et al.*, (2009). They looked at four major tourist destinations in Australia – Kakadu National Park, Wet Tropics in Queensland, Barossa Valley and the Blue Mountains. Detailed climate projections were generated for each of the case studies, providing useful indicators for tourism. For example, rising temperatures in the Northern Territory were predicted to increase the heat stress of tourists and increase operational costs for accommodation providers due to higher air conditioning demands. The frequency of very hot days (above 35°C) in Darwin, for example, currently averages 11 per year, but this is projected to rise to 28–69 days per year by 2030 (mid emissions scenario) and 49–153 days per year by 2070 (low emissions scenario) (CSIRO, 2007). For Richmond in Australia's Blue Mountains, Lucas *et al.* (2007) found that the current annual-average number of very high and extreme fire weather days will increase from 14 at present (1973–2005) to 14–16 by 2020 and 15–24 by 2050. The implications of this increased fire frequency for tourist visitation and image erosion are potentially severe.

The southern Australian heatwaves in January 2009 and 2010 caused tracks to buckle under record 44°C heat, cancelling trains, stranding passengers and delaying goods (International Union of Railways, 2010). For parts of New Zealand where rail tracks are in poor condition the mean number of speed restriction days is projected to double from six per year in 1990 to 13 days per year in 2040 and to 23 days per year by 2090 (Gardiner *et al.*, 2009). In England over £170 million has been allocated to improving operating performance and passenger comfort by increasing the ability of the London Underground cooling system to cope with higher temperatures (Arkell and Darch, 2006). High temperatures have been regularly experienced by passengers in hot weather on the Underground, with reports of temperatures having reached 40°C in one instance. Significantly, hot spells do not result in decreased patronage. An interesting coping response to high temperatures in the Underground has been to distribute bottles of cold mineral water to passengers at selected stations.

Severe weather events and the potential for more intense disasters are increasingly taken seriously by the tourism sector. A recent study by UNEP and CAST

(Caribbean Alliance for Sustainable Tourism) (2008) provides guidelines for the management of disasters in coastal destinations. Other natural disasters of high relevance to tourism are tropical and other storms, floods, tornadoes, and landslides triggered by excessive rain. The 2011 Queensland floods in Australia, which have been attributed to the strongest ever recorded La Niña in the Pacific, had devastating impacts on the tourism industry. While no comprehensive assessments are as yet available, the damage could be several AUS$100 millions (Fenner, 2011). Several tools have been developed to assist the tourism industry in managing the impacts of extreme events, including those developed by Tourism Queensland (2007) and Tourism Victoria (2010). But research has yet to clearly elucidate the linkages between natural disasters, climate change and tourism, and how they might best be managed. Becken *et al.* (2011) provide some indicative findings. Earlier publications, such as the APEC (Asia-Pacific Economic Cooperation) International Centre for Sustainable Tourism (AICST) (2006) guide on tourism risk management, also provide useful information.

> Multiple factors converge to make the people inhabiting coastal zones and small islands highly vulnerable. Coasts and small islands are highly exposed to a variety of climate hazards that may be affected by global climate change. The climatic hazards converge with local and regional human pressures to create conditions of high vulnerability, particularly in areas with high concentrations of people and infrastructure in low lying coasts. (Assessment of Impacts and Adaptation to Climate change, 2007, in UNEP and CAST, 2008: 27)

The ski industry is one of the most vulnerable parts of tourism with respect to climate change, explaining the great research effort in this area (e.g. Agrawala, 2007; Hendrikx and Hreinsson, 2010; Pickering *et al.*, 2009; Scott *et al.*, 2006; Uhlmann *et al.*, 2009). Warming temperatures during winter, coupled with changing precipitation patterns, put increasing pressures on ski fields, especially those at lower elevations (Falk, 2009). Snow models are becoming increasingly sophisticated and are now often commissioned by ski field associations. For the Australian Alps, a 1°C increase in temperature combined with an 8.3 per cent decrease in precipitation will reduce snow duration by about 60 per cent by 2020 (Hennessy *et al.*, 2008). In New Zealand, maximum average snow depths (on 31 August) at ski fields will be reduced to 79–93 per cent of the current maximum average by 2040, and 54–80 per cent by 2090. The specific changes are dependent on elevation and location and on the assumption of a mid-range (A1B) emissions scenario (Hendrikx and Hreinsson, 2010). Interestingly, the implementation of adaptation measures appears to be more related to managers' awareness of climate change than to the actual vulnerability (Hoffmann *et al.*, 2009). This means that is it not necessarily the most vulnerable ski fields that engage in climate change adaptation. Snowmaking will also be limited in places by the availability of water and the costs associated with making snow. Pickering and Buckley (2010) estimated that the required 700 snow guns for Australia's six main resorts would require AUS$100 million in capital investment, and 2500–3300 million litres of water per month.

Empirical research in North America provides evidence that minimum and maximum temperature, snow depth and wind chill are statistically related to the ticket sales for downhill skiing (Shih *et al.*, 2009). Research in Finland showed that high winds were the most common reason for ski field closure, while snowmobiling and cross-country skiing were impeded most by frosty conditions (Tervo, 2008). Impacts on the patronage of ski fields could be significant. In Australia, in snow-poor winters, 69 per cent of skiers would ski less often, 5 per cent would give up and 16 per cent would ski overseas (Pickering *et al.*, 2009). However, an Austrian study on skier demand demonstrated that a skier's income is also a major driver of ski field visitation, in particular for high elevation ski fields where income elasticities reached a value of 1.63 (compared with 0.55 for low elevation ski fields). This finding highlights the need for assessments to go beyond climatic parameters. The literature also indicates that the ski industry has been successful in their adaptation efforts. Increased snowmaking capacity reduces the impacts of warm winters, as shown in an analogue analysis of ski fields in the United States (Dawson *et al.*, 2009). Also, innovative adaptation tools, such as weather derivatives, have recently been explored in the tourism literature (Bank and Wiesner, 2011).

Similar to winter resorts, many beach tourism destinations depend on favourable climatic conditions, for example ample sunshine, no precipitation and no wind (Scott *et al.*, 2008). Warm temperatures, clear waters and low health risks were also found to be the most important environmental features influencing the choice of holiday destination for tourists of two Caribbean islands, Bonaire and Barbados (Uyarra *et al.*, 2005). Research targeting the Great Barrier Reef in Australia indicates that poor weather has a more pronounced effect on satisfaction than does good weather: seasickness, cold or wet conditions, reduced visibility and difficult snorkelling conditions all led to reduced satisfaction levels (Coghlan and Prideaux, 2009). In an attempt to assess the vulnerability of beach tourism destinations to climate change, Perch-Nielsen (2009) developed indicators for 51 countries around the world. While useful at a macro-level, Perch-Nielsen's study fails to take into account local complexities. Examples of more local assessments include the analysis of whale watching tourism by Lambert *et al.* (2010) and the framework for coastal tourism developed by Moreno and Becken (2009). A study of beach visitation in East Anglia, England, for example, showed that sea-level rise would impact on beaches by reducing the width of the beach. However, resulting potential reductions in visitation were found to be outweighed by increased visitation due to more favourable temperatures (Coombers *et al.*, 2009). These in turn will increase pressure on local biodiversity, in particular vegetation cover and habitat for nesting birds such as the ringed plover (Coombers *et al.*, 2008).

The consequences of climate change for tourist destinations are far reaching, as they affect the resource base of tourism (e.g. see UNESCO, 2007 on World Heritage Areas), both directly and indirectly (McEvoy *et al.*, 2006; Tratalos *et al.*, 2005). As such there may well be 'winners and losers' as suggested in a recent Deutsche Bank report (Ehmer and Heymann, 2008). Importantly, a small number of studies are beginning to explore opportunities associated with climate change. The golf industry in Canada, for example, is projected to be positively affected by climate change because warmer temperatures will alter the length of the operating season and influence seasonal and total golf demand. Golf courses on Canada's east coast could witness an increase in operating season by 25 to 45 days by 2020 (Scott and Jones, 2007). On the other hand, drier conditions will require increased turf management as a result of a greater need for irrigation, turf grass selection and turf disease and pest management (UNWTO *et al.*, 2008). Similarly, inhabitants of the Baltic coast region are becoming more aware of the new opportunities offered by climate change (Schernewski, 2009). Potential growth opportunities for tourism in the Arctic have been explored in the light of melting sea ice in the northern hemisphere and a greater accessibility for Arctic cruise tourism. However, Stewart *et al.* (2007) note that this might be a double-edged sword. While ships might be able to navigate to previously inaccessible areas, the environmental conditions could reduce the attractiveness of destinations as a result of changing landscapes and reduced wildlife viewing opportunities. There will also be areas (e.g. the North-west Passage) where warmer conditions will increase sea ice hazards that pose safety concerns for tourism.

2.9 Future-oriented analyses

Some recent studies have focused on the future of tourism, with particular consideration of climate change effects. A major, industry-led initiative in the United Kingdom explored four different scenarios for the future of tourism. In these scenarios, both climate change and energy constraints were seen as the key defining dimensions of the future of tourism (Draper *et al.*, 2009). The resulting analyses point to increasing transport costs due to rising oil prices and carbon policies. This differs markedly from the majority of tourism future analyses which assume that the model of low-cost carriers will increase its market penetration and lead to more affordable travel (Yeoman *et al.*, 2009). The previously mentioned UK 'Tourism 2023' study (Draper *et al.*, 2009) stands out as the only investigation which distinguished between trends that are both more and less certain. Climate change impacts, water scarcity, cost of resources (including oil), growth in visitor numbers, ageing population and political instability were the more certain factors. Among others, how environmental attitudes affect mobility, future legislation and how expensive oil would be were perceived as less certain (Draper *et al.*, 2009). Several researchers have voiced concerns about the future growth of tourism and argued that 'radical change will be needed to reconcile the holiday and business travel demands of a growing world population with the climate policy targets of the international community' (Gössling *et al.*, 2010a: 119).

Characterizing the future in a carbon-constrained world is difficult, forcing researchers to employ a variety of methods. Informed fiction writing, such as Urry's (2008) future scenario of a 'digital panopticum' is one possible approach. In this scenario, Urry envisages a new 'architecture of technologies and practices' (p. 270) that emerges in the form of smart and digitized transport systems that provide an alternative to the automobile-driven society of today. Backcasting is another possible method of understanding the link between the present and the future. Dubois *et al.* (2011) assess the future tourism mobility of the world population and conclude that GHG emissions that are 50–80 per cent less than present levels can be achieved with only significant changes in transportation systems in combination with a decrease in distances travelled. Air transport will have to grow slower, stagnate or even decline. The authors also point out that current travel patterns in industrialized countries would have to change. This can only be achieved through demand management, such as the use of pricing mechanisms.

Predicting the economic impact of climate change on tourism is incredibly challenging. This is largely due to the difficulty in understanding tourism demand and behaviours for time frames that are well beyond normal tourism planning horizons. Climate models can provide projections for 50 or 100 years, but these are not easily aligned with tourism analyses. An exploratory study to quantify the potential economic impacts on five selected Australian tourism destinations (Pham *et al.*, 2010) found that reduced demand due to increasingly unfavourable climatic conditions adds to other negative impacts at regional level, such as declining agricultural yields. While the economic impact of climate change on tourism at these five destinations was relatively insignificant at the national level, at the regional level it was considerable, but varied widely between regions. In a similar study of tourism in the Bernese Oberland (Switzerland) in 2030, it was estimated (based on climate models in combination with expert assessments) that tourism revenue would decline by 4 per cent, including the costs of implementing adaptation measures. However, adaptation interventions, such as greater investment in summer tourism, would substantially offset climate change induced losses (Müller and Weber, 2008).

The future cost of climate change to the Caribbean tourism sector has been estimated by combining the impacts of reduced tourist arrivals (a consequence of new climate policies and changes in weather patterns, with the latter being estimated using a tourism climatic index), loss of coral reefs and adverse impacts due to sea-level rise and extreme events (ECLAC, 2011). The findings suggest that the impact of climate change on future tourism demand (arrivals) and the Caribbean tourism product will lead to severe decreases in industry earnings and in government revenues. Due to their low levels of current and future GHG emissions, countries in the Caribbean have few options to lessen their emissions in the future. On the other hand, the study concluded that the significant impacts of climate change for Caribbean countries in the future means that adaptation to climate change must be viewed not just as a way to reduce future risks but also as an imperative to ensure their continuing economic viability.

These and similar findings in other studies (e.g. Matasci and Altamirano-Cabrera, 2010) provide support for the importance of investment in adaptation to reduce the future impacts of climate change, and facilitate sustainable tourism development in the longer term.

2.10 Conclusion

Research on climate change and tourism has increased substantially in the last five years. The emphasis is still on tourism's GHG emissions and 'carbon footprints', but an increasing number of studies are assessing the impacts of climate change on tourism and identifying potential adaptation options. Overall, the body of knowledge is biased towards developed countries, particularly in Europe, Oceania and, to a lesser extent, in North America. The Caribbean is the only 'non-Western' region that has invested substantially in research and knowledge generation to address the implications of climate change for the tourism sector. Very little is known about the situation for destinations in Africa, South America and even Asia, despite their growing importance as global tourism players. The latest insights from climate change science highlight the urgency with which climate change must be addressed in order to prepare for potentially significant changes. An increase in extreme events will have especially important repercussions for tourism destinations and operators. The preceding synthesis of our knowledge and experience also highlights that tourism stakeholders around the world have yet to make much progress in developing and implementing the policies and plans required to face these challenges.

3

ADAPTATION POLICY, PLANNING AND INSTITUTIONS

3.1 Introduction

The substantial, multi-faceted and far-reaching consequences of climate change for tourism necessitate a coordinated, in-depth response that takes a proactive approach. In some cases this will entail preparation of a policy and plan of action. It may even involve establishing new, or strengthening existing, institutional arrangements. Such comprehensive responses are typically undertaken by national and local governments, usually in situations where tourism makes a significant contribution to the economy and to the lives and livelihoods of the population. Large tourism corporations and tourism industry organizations also take such initiatives, in order to reduce the risks that climate change represents to the productivity and sustainability of the business, or the industry as a whole. In many cases these initiatives will also seek to exploit any benefits climate change might bring to the tourism industry.

All these initiatives are part of strengthening the enabling environment for adaptation. As such they represent preparatory activities designed to ensure that practical measures to reduce the consequences of climate change for those engaged in tourism will be both efficient and effective. As Figure 3.1 shows, this strengthening of the enabling environment is a desirable first step in the adaptation process, or cycle. Figure 3.1 also shows that the enabling environment encompasses more than policies, plans and institutions, but these are the focus of the present chapter.

Smaller tourism operators, and tourists themselves, should benefit from these higher-level policies, plans and institutional arrangements. As a result, they are often consulted when policies and plans are being developed. Smaller operators might well prepare an adaptation plan as part of their wider business plan, especially if the higher-level policies and plans indicate their business is at significant risk. Climate change policies and plans for the tourism sector often cover both mitigation (see Chapter 5) and adaptation, with the latter also including disaster risk reduction

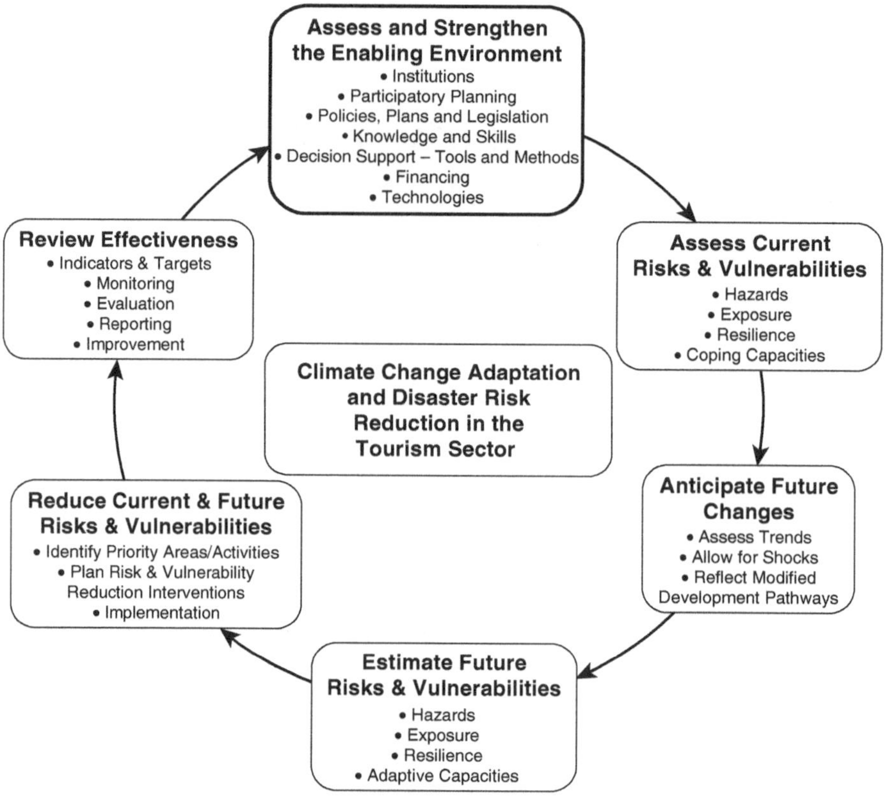

FIGURE 3.1 The adaptation process, highlighting the importance of a strong enabling environment for effective adaptation in the tourism sector.

(Figure 3.1). This is because there are many synergistic benefits of addressing all three together, as is explored further in Chapter 8.

In a recent report, the OECD and UNEP (2011) analyse policies and issues related to climate change adaptation in the tourism sector, to inform the identification of priority areas to be included in a framework for action in the area of climate change and tourism. A key conclusion is that few countries have reported on comprehensive analyses of impacts, adaptive capacities and associated vulnerabilities. There are only a limited number of specific policies addressing adaptation in tourism, even though legislation and policies exist in some countries to deal with relevant issues such as flooding and water. Often these are not specifically related to climate change or tourism.

While the progress made to date reflects growing awareness of the implications of climate change for tourism, important knowledge and skill gaps are yet to be addressed. A key concern is that, while some adaptation assessments focus on identifying and managing impacts that might occur more than half a century from now, it is unclear how policy makers, planners and tourism operators can best address such long-term issues. A significant conclusion reached in the joint OECD and UNEP report is that climate change adaptation assessments related to tourism as well as other sectors need to integrate greater complexity, relevant timescales, as well as the levels (tourist, business, city, region, country) at which policies can be implemented.

As is common with all policies, an adaptation policy for tourism will normally include a vision and goal, describing longer- and nearer-term ambitions for managing the increasing risks to tourism as a result of climate change. Typically the policy will also describe shorter-term objectives to achieve these ambitions, as well as principles that underpin both the development of the policy and its implementation. The latter is usually guided by an implementation or action plan. The principles might include the notions of actions to reduce climate risks for the tourism sector being: (i) relevant, appropriate and proportional to the level of risk; (ii) collaborative and coordinated; (iii) facilitated and guided by traditional knowledge; (iv) integrated, with synergies exploited; (v) measurable; and (vi) flexible, as new knowledge and experience comes to hand.

Adaptation planning, including planning for disaster risk reduction, builds on the adaptation policy by creating a practical document that takes each policy objective and elaborates it in terms of the anticipated results. These are generally described in terms of planned outputs and intended outcomes. A plan will also describe the specific activities that will be undertaken to deliver the results, as well as who will undertake the actions, how the activities will be managed and coordinated, the financial and other resource requirements, and a timetable for implementation including significant milestones. Finally, a plan will usually describe how implementation will be monitored and evaluated, including the performance targets and the indicators and sources of information by which this performance will be measured and reported.

The following sections provide, and comment on, examples of both climate change policies and plans prepared by the tourism sector, covering local, national, multi-country and international levels. Reference is also made to institutional arrangements, as these are relevant to the preparation and implementation of adaptation policies and plans by key players in the tourism sector. Tourism and climate change involve interactions that extend way beyond the sector itself. Thus coordination of activities to address the impact of climate change on tourism is of the utmost importance. Strong and effective institutions in both the public and private sectors are critical to successful adaptation. Dieke (2005) identifies institutional structures as one of the critical success factors for tourism policy, planning, regulatory frameworks and implementation. In many countries too much effort is given to the technical aspects of tourism planning, and not enough to the ability of the existing institutional arrangements to implement planned activities.

3.2 Ensuring a strong enabling environment

As noted above, a strong enabling environment for adaptation is critical to the success of any adaptation initiative. The enabling environment consists of the institutions, policies, plans and legislation, knowledge and skills, decision support tools and methods, financing and technologies (Figure 3.1). Policies and plans to manage the adverse consequences of climate change for tourism, and exploit any benefits, should pay close attention to the enabling environment, in terms of both its current status as well as any need for it to be strengthened.

Case study: New Zealand Tourism and Climate Change Plan

Two of the four priority areas and associated actions in the New Zealand Tourism and Climate Change Plan (New Zealand Ministry of Tourism, 2008) are related to strengthening three key aspects of the enabling environment, namely decision support, legislation/regulation and knowledge (see Table 3.1). In addition, the Plan allocates responsibilities to specific agencies or other institutions, identifies resource requirements and specifies the timing for implementation.

Importantly, the Plan would likely be more effective had it made explicit reference to other key components of the enabling environment, such as institutional strengthening and mobilization of new financial resources. It also fails to describe how implementation of the Plan will be monitored and evaluated against performance targets and indicators. Additionally, in the Plan actions to address climate change are largely targeted at mitigation, and often at the perceptions tourists have of the need for mitigation. The Plan underestimates the need for adaptation, with the only actions that specifically deal with adaptation being research related. This relates to a current research project that investigates the climate change impacts on

TABLE 3.1 New Zealand Tourism and Climate Change Plan: priority areas and actions related to strengthening the enabling environment. (New Zealand Ministry of Tourism, 2008)

Priority area: enable and ensure an ongoing response by tourism businesses	*Priority area: establish a forward research agenda for climate change and tourism*
Examples of specific actions: • Enabling businesses to make good decisions on their response to climate change through information sharing and education. • Align and enhance existing tourism environmental sustainability levers to include climate change. • Address environmentally damaging tourism sector practices through regulatory measures when these practices pose unacceptable risks to tourism and cannot be resolved by other means.	*Examples of specific actions*: • Prioritise research within existing science and tourism funding channels to identify physical impacts on New Zealand tourism from climate change effects for likely warming scenarios over time. • In parallel with physical impact research, identify adaptation issues, options, timings and costs that could be implemented by the sector to mitigate the physical impacts of climate change. • Convene a process to identify the options and implications for New Zealand tourism should climate change concerns become significantly more acute.

New Zealand tourism, including a focus on extreme weather events and the provision of climate information to tourists and other tourism stakeholders (Plate 3.1).

3.3 Incorporating managing climate risks into national policies and plans

The previous section has provided an example of a standalone policy and planning instrument related to tourism and climate change. Through the use of two case studies, for Saint Lucia and Jamaica, this section illustrates how climate risks to tourism can be addressed in policy making and planning as part of more comprehensive

PLATE 3.1 Information for tourists on road conditions after a heavy snowfall event in Wanaka, New Zealand. (J. Wilson)

national climate change policies or plans that include tourism by way of either a sector-by-sector approach or when addressing issues that are of direct but more marginal importance to tourism, such as water resource management and biodiversity protection.

Case study of tourism as a sector in a national adaptation policy

Saint Lucia is a small island country in the Caribbean with a population of less than 200,000. It attracts over 350,000 tourists a year, due in part to its tropical climate, scenery and large number of beaches. As a result, tourism makes up more than half of the country's GDP. Saint Lucia's National Climate Change Adaptation Policy and Strategy (Government of Saint Lucia, no date) addresses climate risks to tourism, along with those for seven other sectors. The climate risks related to tourism include damage to and destruction of hotels and other tourism infrastructure located in coastal areas susceptible to storm surges, erosion and sea-level rise; loss of economic returns due to the possible changes in, or loss of, coral reefs, beaches, natural forests and other natural resources and attractions; reduced visitor arrivals as a result of a higher frequency of extreme weather events such as hurricanes, as well as reduced inducement for travel as a result of higher temperatures in traditional tourism

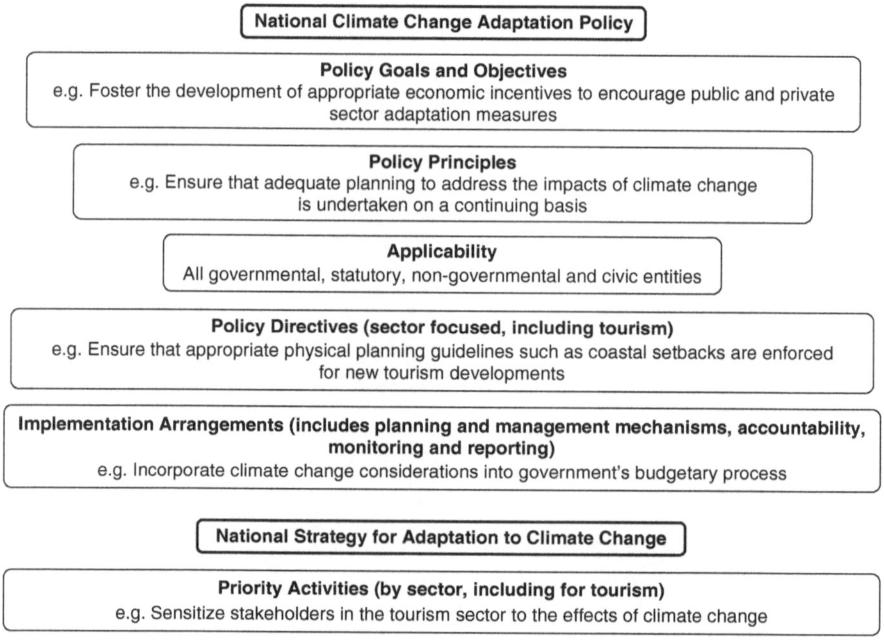

FIGURE 3.2 Structure and content of Saint Lucia's National Climate Change Adaptation Policy and Strategy.

centres; and negative changes in water and food arising from changes in precipitation levels and distribution, loss of forest cover and related factors.

As shown in Figure 3.2, the Policy conforms with good practice by specifying goals, objectives, principles, planning and management mechanisms, accountability, monitoring and review. In an effort to ensure appropriate approaches to adaptation in the tourism sector the Policy identifies the following activities: (i) conduct the necessary research and information gathering in order to strengthen the basis for sound decision making; (ii) ensure that appropriate physical planning guidelines, such as coastal setbacks, are enforced for new tourism developments; and (iii) work with stakeholders in the tourism sector to develop a strategic plan which incorporates climate change considerations and appropriate measures such as water conservation programmes as well as general sustainability concerns.

Thus, with respect to tourism, the Policy has a focus on strengthening the enabling environment for adaptation. This is also the case for Saint Lucia's National Strategy for Adaptation to Climate Change. As shown in Figure 3.2, the Strategy is a direct derivative of the Climate Change Adaptation Policy. As such it is intended to serve as a blueprint for implementation of the Policy. The Strategy is divided into a number of components, each of which corresponds to a sector under the Policy, including tourism. The strategies and actions related to tourism are shown in Table 3.2. Also included are the time frame and responsibilities for implementation, and resource requirements. Therefore the Policy and associated Strategy have all the content one would expect to find in such policy and planning documents,

TABLE 3.2 Saint Lucia's National Strategy for Adaptation to Climate Change by the Tourism Sector. (After Government of Saint Lucia, no date)

Strategy & action	*Rationale*	*Priority*	*Time frame*	*Resource needs*
Sensitize stakeholders in the tourism industry to the effects of climate change	To influence the policy making process in order to achieve climate change sensitive decision making	Medium	Short and long term	Relevant climate data, human resources
Improve and develop framework with emphasis on enforcement	Make tourism activities adaptive to climate change through enforcement	High	Long term	Human resources
Development and implementation of integrated, sustained and coordinated public education and awareness programmes	To sensitize the public as to the impacts of climate change	High	Long term	Human, technical, financial training

but the exclusive focus on strengthening the enabling environment for adaptation, rather than on-the-ground implementation, means there will be little tangible impact on reducing climate-related risks to the tourism sector.

Case study: National policy addressing issues of direct or more marginal relevance to tourism

Often the third approach to managing climate risks is referred to as 'mainstreaming' or 'integrating' climate change considerations into national or sector policy and planning processes. In some instances the specific national policy or plan may have only marginal relevance to tourism. As the following case study will illustrate, the emphasis in this approach is on ensuring that climate risks to tourism are managed in an integral manner, as part of the work being undertaken nationally or at the sector level, rather than in a more standalone manner, as is the situation for the previous two case studies.

Jamaica is one of the best-known resort vacation destinations in the world (Plate 3.2). Since the eighteenth century it has been a favoured retreat for travellers from Europe. In 2006, total arrivals exceeded three million. Tourism is now one of the leading industries in Jamaica, contributing an estimated 8 per cent of total GDP, directly employing approximately 80,000 persons and accounting for some

PLATE 3.2 Tourist beach-scene in Ocho Rios, Jamaica, on a windy day. (R. Mahon)

47 per cent of total foreign exchange earnings from productive sectors (Jamaica Tourism Task Force, 2009).

Significantly, Jamaica's Master Plan for Sustainable Tourism Development (2001–2010) did not refer to climate change. However, Vision 2030 Jamaica (Government of Jamaica, 2009), the national development plan, provides a framework designed to ensure that climate change issues are mainstreamed into national policies and development activities. In this way, the plan recognizes that there are inadequate policy and planning linkages between tourism and the other sectors of the economy, including agriculture, manufacturing and services. One objective of Vision 2030 is to avoid or minimize the impact of climate change related disasters through increased coping capacity at various levels within the country, such as economic sectors and communities (Plate 3.3). This includes developing and applying multi-hazard risk reduction measures for the tourism sector.

A key area for action proposed in Vision 2030 is to embed both climate change mitigation and adaptation in the physical planning system. It recognizes that, to achieve significant reductions in climate risk, adaptation measures need to take into account challenges faced by the key economic sectors of agriculture and tourism, including location of social and economic infrastructure, livelihoods, location of settlements, biodiversity, water production, land use and management. Another example of a key area for action is to ensure that, since natural ecosystems are an

PLATE 3.3 Damage caused to the Barcelo Maya Resort, Quintara Roo, Mexico, as a result of Hurricane Wilma. (J. Kloss)

important part of tourism's resource base, the activities of the tourism sector must support biodiversity conservation objectives. All of the proposed actions highlight the benefits of incorporating the management of climate risks to tourism in wider national policies and plans. One shortcoming of the policy is that all the outcome indicators and proposed targets for the tourism sector relate to its economic performance and international competiveness. Since one objective of Vision 2030 is to avoid or minimize the impact of climate change related disasters it would be appropriate to include an indicator that directly reflects the risks faced by the tourism sector. With this approach, performance evaluation does not have to wait until an extreme event occurs. By then it is too late to realize that an adaptation intervention has been less than effective.

Other examples of national adaptation plans and policies

A number of (largely developed) countries have prepared a policy or strategy to address climate change and tourism, with various approaches being used. Some prominent examples are compiled below.

Country	*Description*	*Source*
France	ONERC: Stratégie nationale d'adaptation. Recognizes that climate change will create the need to diversity tourism offerings and may even lead to closure of some operations and opening of others.	www.developpement-durable.gouv.fr/IMG/pdf/Strategie_Nationale_2-17_Mo-2-2.pdf
Tunisia	A number of projects have been completed or are still being implemented. A national adaptation strategy was supported by GTZ including foci on agriculture, tourism and health.	www.gtz.de/en/dokumente/en-brochure-unfccc.pdf and http://www.tn.undp.org
United Kingdom	Defra's Climate Change Plan recognizes the opportunities for increased domestic tourism and challenges related to changes in water availability.	Department for Environment, Food and Rural Affairs (2010)
Germany	The German Adaptation Strategy creates a framework for adaptation to the consequences of climate change. Water regime and management, coastal and marine protection and tourism are identified as important areas for action. Rising temperatures at lower altitudes will lead to a concentration of ski tourism in the higher, more ecologically sensitive regions.	www.bmu.de/files/english/pdf/application/pdf/das_gesamt_en_bf.pdf

Country	*Description*	*Source*
Mexico	The National Climate Change Strategy (NCCS) identifies opportunities for adaptation in specific sectors, including tourism. The Special Climate Change Programme (SCCP) identifies the areas that are vulnerable to the effects of climate change. Regional climate forecasts and early warning systems will be established for eight important tourism destinations.	Government of Mexico (2007, 2009)
Sweden	Recognizes opportunities for increased tourism in a changed climate, with warmer summers and higher bathing water temperatures. However, water resources and quality will remain a key issue. Winter tourism and outdoor leisure will be affected by 'poor' winters, and adaptation measures will be required.	Sweden facing climate change – threats and opportunities (www.regeringen.se/content/1/c6/09/60/02/56302ee7.pdf)
Scotland	The Economic Strategy outlines how seven key sectors, including tourism, will be impacted by climate change, with new opportunities for businesses as a result of increased demand for certain services and products and opportunities for innovation, including those that stem from regulatory changes.	www.scotland.gov.uk/Publications/2009/12/08130835/1
Samoa	In the Samoa Tourism Development Plan (2009–2013), climate change is considered an extreme risk; the response strategy is to ensure accurate identification of potential changes in the physical environment due to climate change and to factor these into planning and development decisions; but there are no proposed actions to address these risks.	Samoa Tourism Authority (2009)

3.4 Adaptation policy making and planning at the sub-national and business levels

Local government, local tourism industry organizations and individual operators also have major roles in managing the climate risks to tourism. While they can be guided by national, regional and even international policies and other initiatives, the local and often site-specific nature of climate change impacts means that context-focused policies and plans will provide more useful guidance and coordination for tourism operators and those who provide services and support.

Ribeiro *et al.* (2009) provide policy guidance intended to assist sub-national authorities and associated bodies to prepare regional (i.e. sub-national) adaptation strategies. The guidance draws on the experience of 31 regional adaptation strategies, a review of existing guidelines for developing adaptation strategies, and on the results of a stakeholder's workshop to discuss a draft of the guidance. Several key findings are presented, notably that:

- The strategies differ in their geographical scope as some countries are planned at municipal level while in other countries planning is at regional (i.e. sub-national) level.
- National adaptation strategies often provide a framework for the development of regional strategies.
- Existing regional adaptation strategies tend to focus on a patchwork of scientific background information and methods, are based mostly on climate variation impacts and less on adaptive capacity, and seldom have their strategic assessments undertaken.
- Often little attention is paid to implementation issues, such as financing.

Case study of adaptation policy making and planning at destination level

The Sunshine Coast, on the eastern coast of Australia, is a centre for tourism as a result of its extensive surf beaches, mild weather, theme parks, national parks and numerous golf courses. Tourism is one of the three major industry sectors on the Sunshine Coast, attracting approximately three million visitors each year. As a result, the Sunshine Coast also features as a case study for local mitigation policies (Chapter 5). Tourism and related industries such as retail and hospitality on the Sunshine Coast are particularly exposed to the direct and indirect impacts of climate change. Coastal amenity and lifestyle attractions are key to the region's popularity. But these are immediately affected by climate change and associated extreme weather events.

The area has been identified by the IPCC as a climate change 'hotspot'. One aspect of climate change that is of great concern to the Sunshine Coast, as elsewhere, is that, in addition to sea-level rise, climate change alters the frequency of high sea-level extremes such as storm surges and king tides (Sunshine Coast Council, 2010a). Generally, for the southeast Queensland region, even a reasonably modest rise of sea level of 0.5 m will cause the frequency of high sea-level events to likely multiply by 10,000. This means that the present 100-year event would happen several times a year (Sunshine Coast Council, 2010b). On the Sunshine Coast, higher storm surges and larger spring tides will result in an increased risk of coastal erosion and flooding. Such events are already being experienced with increasing frequency and severity (Plate 3.4).

In addition to higher sea levels, the increase in sea-level extremes is also a result of a projected increase in storm tides and wave set up. While a 9 per cent decrease

PLATE 3.4 Australia's Gold Coast, the recently announced host city for the 2018 Commonwealth Games. (J. E. Hay)

in tropical cyclone numbers is projected, an increase in the number of long-lived tropical cyclones is anticipated, as are storms with lower central pressures. A 60 per cent increase in the number of the most severe storms is likely by 2030, with a 140 per cent increase by 2070 (Sunshine Coast Council, 2010b). These changes will be reflected in stronger winds and hence greater storm tides and wave set up.

In response to these threats, and those related to a lesser-known threat of declining oil supplies (see Chapter 8), the Sunshine Coast Council prepared the Sunshine Coast Climate Change and Peak Oil Strategy (2010–2020), and an associated Action Plan. The Strategy and Plan recognize that exposure to climate change is heightened by the region's coastal location, rapid population growth and associated development pressures, the dispersed settlement pattern and reliance on climate-sensitive economic activities. A strategic framework has also been developed to support the implementation of the Strategy and Action Plan to achieve the overall goal of building and maintaining environmental, social and economic resilience to the two challenges of climate change and peak oil. The Strategy and Action Plan recognize that some of the actions required to address the challenge of responding to climate change fall outside the responsibility of the Council, and are thus likely to also require policy responses from other levels of government, as well as civil society (including the private sector).

While many of the actions identified in the Action Plan will be undertaken using existing resources and budget allocations, in other cases actions will be subject to the Council's annual budget process or may be funded through the Environment Levy. The Action Plan uses cost indicators, where 'low cost' is AUS$0–$100,000, 'medium cost' is AUS$100,000–$1,000,000, and 'high cost' is greater than AUS$1,000,000. The cost of action indicates up-front costs only and does not factor in the return on investment. It is important to note that, in several cases, implementing the activities will result in significant long-term financial savings. The Action Plan also sets out the time frame for the actions: immediate actions to be undertaken within the first year; short-term actions to be undertaken within three years; medium-term actions to be undertaken within six years; and long-term actions to be undertaken after six years. There is also a category of ongoing action. Table 3.3 provides examples of adaptation interventions in the Action Plan that are of relevance to the tourism sector.

The Council will regularly report on progress in implementing the Strategy, including annual reporting of key performance indicators that relate to targets in the Council's Corporate Plan. The Strategy will be reviewed every three years to respond to emerging issues. A more formal review and update of the Strategy and its supporting plans will be undertaken after five years, to reflect developments in science, technology and government policy directions. Thus, in combination, the Strategy and Action Plan cover all the components of the adaptation process (Figure 3.1), moving from strengthening the enabling environment through to monitoring, evaluation and reporting. Collectively, they also address policy goals through to

TABLE 3.3 Examples of adaptation actions of relevance to the tourism sector. (Sunshine Coast Council, 2010a)

Action	*Time frame*	*Cost*
Undertake initial vulnerability and hazard mapping to identify major risk areas due to climate change on the Sunshine Coast	Immediate	Medium
Incorporate climate change into all hydrological mapping and forecasting	Immediate	Low
Complete climate change risk assessments	Medium	Medium
Using planning approaches, protect, enhance and connect strategic wildlife corridors and riparian corridors, to allow for species shift, and reduce pest species	Immediate and ongoing	High
Encourage water harvesting, water reuse and other water reduction strategies in planning scheme provisions	Medium	Low
Promote the potential economic benefits that may be derived through adaptation to accommodate wintering tourists and better growing opportunities	Medium	Low

actions, costs and time frames. However, performance targets and indicators are not included in the Strategy and its Plan, even though the Strategy refers to annual reporting of key performance indicators that relate to targets in the Council's Corporate Plan. This omission will, to some extent, compromise the monitoring, evaluation and reporting procedures and resulting learning.

Case study of climate adaptation for tourism in a national park

Kakadu National Park is located 200 kilometres east of Darwin in the wet-dry tropics of the Northern Territory. It covers an area of 19,798 square kilometres. Currently, about 210,000 people visit Kakadu each year. Kakadu is one of the very few places listed as a World Heritage Area for both its cultural and natural values. The Kakadu National Park Climate Change Strategy 2010–2015 (Government of Australia, 2010) recommends the preliminary adaptation, mitigation and communication actions that are required to manage the consequences of climate change and reduce the carbon footprint of the park. The Strategy is consistent with the policies and actions of the Kakadu National Park Management Plan 2007–2014 and the objectives identified in the Parks Australia Climate Change Strategic Overview 2009–2014. The Strategy is an adaptive tool subject to ongoing review, with management responses being amended to take account of improvements in the understanding of the implications of climate change for the Park. The Strategy will be reviewed on a rolling basis, to also take account of new information or changes in policy directions.

The Strategy recognizes that, while average rainfall is not expected to change significantly in the coming decades, evaporation is expected to increase substantially, with more extended dry periods. There is likely to be an increase in extreme weather events such as Category 5 tropical cyclones and more frequent and intense rainfall events. The number of days with temperatures exceeding 35°C is anticipated to increase from the current 11 days per year to over 295 days per year by 2070. By this time sea level is expected to have risen by 50 cm. The key consequences of climate change to the Park are considered to be saltwater incursions into freshwater ecosystems, changing fire seasons, potential for spread of exotic flora and fauna, more intense storm activity, loss of sacred sites, damage to rock art sites and challenges to the sustainability of some indigenous communities located in low-lying coastal areas. Increased atmospheric CO_2 concentrations may increase photosynthesis and plant biomass of species that impact on vegetation community structure and composition. More extreme climatic conditions are expected to place additional pressure on the resilience and suitability of park infrastructure. This is likely to have flow-on implications for maintenance costs. Higher temperatures are likely to increase the demand for energy and water for cooling systems. Infrastructure to meet this growth in demand for peak generating capacity needs to be managed to ensure that the Park's carbon footprint is minimized as far as practicable.

There may be a compression of the tourist season due to increased flooding and higher daytime temperatures. The anticipated increase in days with temperatures exceeding 35°C, and changing rainfall patterns, are likely to impact on visitor and staff comfort and satisfaction. Saltwater intrusion of freshwater wetlands may also impact on visitor expectations of the Park. The relationship between visitor expectations and actual visitor experiences in relation to fauna and flora interactions, park access and weather conditions will need to be managed as the impact of climate change becomes better understood. The incidence of heat stroke and heat stress is also likely to increase. More extreme weather events may increase the risk of injury to both park visitors and staff. Changes in temperature and rainfall patterns may lead to an increase in the transmission of diseases by insects, posing significant challenges to the management of human health, particularly in indigenous communities that are disproportionately vulnerable to climate change. On the other hand, climate change could offer some economic benefits to indigenous communities through an increase in 'last chance tourism', participation in carbon trading programmes and employment opportunities in monitoring the impacts of climate change and undertaking remedial and mitigation activities.

With reference to adaptation, the Strategy has four broad objectives. These, and examples of adaptation interventions and time frames, are listed in Table 3.4.

TABLE 3.4 Examples of specific adaptation actions and time frames included in the Kakadu National Park Climate Change Strategy 2010–2015. (Government of Australia, 2010)

Objective	*Examples of adaptation interventions*	*Time frame*
Understand the implications of climate change	Identify baseline parameters to effectively monitor the effects of climate change	By 2012
	Continue to improve data management and the technical capacity of park staff	Ongoing
Implement adaptation measures to maximise the resilience of the reserves	Identify species and habitats where impacts from feral pests, weeds and fire are likely to be exacerbated by climate change	By 2012
	Obtain expert engineering and environmental advice on measures needed to protect significant freshwater habitats from saltwater intrusion	By 2015
Work with communities, industries and stakeholders to adapt to climate change	Investigate opportunities to assist tourism businesses dependent on the park to adapt their businesses to the impacts of climate change	Ongoing
	Continue the park's active role on the local Counter Disaster Committee and continue to hold joint exercises	Ongoing

TABLE 3.4 (*Cont'd*)

Objective	*Examples of adaptation interventions*	*Time frame*
	Conduct an infrastructure risk assessment to identify assets at risk from climate change impacts and extreme weather events.	By 2012
Communicate the implications of, and management responses to, climate change	Develop and implement a communications strategy to better inform stakeholders about the implications of, and the management response to, climate change	By 2012
	Support the maintenance of publicly available information on climate change policies, strategies and other documents relevant to the Park	Ongoing

Case study: Nolitours Hurricane Policy

The Canadian company, Nolitours, offers a comprehensive range of vacation packages to destinations such as Cuba, Mexico, the Dominican Republic, Costa Rica and Nicaragua. Nolitours' Hurricane Policy provides protection to travellers

PLATE 3.5 A (former) restaurant destroyed by Category 5 Hurricane Wilma in October 2005, Cozumel, Mexico. (S. Becken)

should a hurricane affect their destination city in Mexico (Plate 3.5), the Caribbean or Central America. The policy, which does not apply to cruise packages, covers departures between 1 June and 30 November. The insurance cover is activated if either: (i) the Government of Canada issues a travel advisory urging travellers to avoid the destination city because of a hurricane and the US National Hurricane Centre/Tropical Prediction Centre (NHC/TPC) issues a hurricane advisory for the destination city; or (ii) a hurricane designated as a Category 1 Hurricane or higher on the Saffir-Simpson Scale used by the NHC/TPC has already passed through the selected hotel complex within one week of the scheduled departure date.

If the above conditions are met clients may choose to travel at a later date, travel to another Nolitours destination city that is not at risk of a hurricane, or cancel, with a full refund except for any travel insurance premiums already paid. In the event of a client being repatriated to Canada, or being evacuated from a destination city due to a hurricane more than 48 hours after their arrival at a destination while on a Nolitours vacation, the client is eligible for a credit voucher (valid for one year) for unused days on the accommodation portion of the package. However, costs related to the air and land transfer portions of the package are not refundable, and any additional expenses incurred as a result of the hurricane will be borne by the client. Passengers travelling 'air only' will not be eligible for a refund for their accommodation or land transportation costs. If repatriation takes place less than 48 hours after departure, passengers will be eligible for a credit voucher equivalent to the total value of the purchased package.

Other examples of local adaptation policies in the tourism sector

While not so common as their practical engagement in adaptation, there is some evidence that tourist destinations and businesses are beginning to become involved in policy making for climate change adaptation. Policies include any kind of strategic approach or document that seeks to prepare for future climatic changes.

Policy	*Description*	*Source*
Blue Haven, Tobago	The environmental brochure for the Blue Haven Hotel includes actions that achieve a 50 per cent reduction in the use of potable water; the initiatives include a state-of-the-art sewage treatment plant, rainwater collection, irrigation using effluent water and rainwater, and water-saving flush toilets, showers and taps.	www.bluehavenhotel.com/
Ski Areas, North America	Sustainable Slopes Environmental Charter: integrative across water conservation, environmental management, wildlife and conservation issues, and greenhouse gas emissions. Similar policies have been developed in Australia and New Zealand.	www.nsaa.org/nsaa/environment/sustainable_slopes/Charter.pdf

Policy	*Description*	*Source*
Climate Change and the Visitor Economy, England's Northwest	Discusses how the visitor economy can realize the opportunities presented by climate change, while ensuring that the resource base is sustained under growing visitor demand and climate related reductions in environmental capacity.	www.ukcip.org.uk/wordpress/wpcontent/PDFs/CCVE_NW_tech.pdf
Ontario, Canada	Implementing pilot adaptation initiatives in the tourism sector, including encouraging actions to help build climate resiliency through initiatives to expand summer tourism as a result of the warm-weather tourism season extending earlier in the spring and later in the fall and the shortening of the winter recreation season, including decreased reliability of suitable snow cover.	Climate Ready: Ontario's Adaptation Strategy and Action Plan, 2011–2011 (Government of Ontario, 2011)
Great Barrier Reef, Australia	Great Barrier Reef Tourism Climate Change Action Strategy 2009–2012; Great Barrier Reef Tourism Climate Change Action Strategy; Climate Change Action Responsible Reef Practices; Information on Climate Change on the Great Barrier Reef; Reef Facts for Tour Guides.	Zeppel (2011)
'Towards 2015' regional tourism strategy, England	Launched by South West Tourism, the regional tourist board for the South West of England. Three 'Year of Action' plans focus on a particular resource issue and the management of climate change impacts. This involves creating flood risk assessment plans and adapting camping and caravan sites to reflect changing climatic conditions.	www.swtourismalliance.org.uk/
Phuket, Thailand	Phuket Tourism Risk Management Strategy, developed in response to the 2004 tsunami, but also applicable to wider disaster risk reduction and response to extreme climatic events. Based on risk management approach and stakeholder inclusive.	AICST (2006)
London, England	The aim of the London Climate Change Adaptation Strategy is to assess the consequences of climate change on London and to prepare for the impacts of climate change and extreme weather to protect and enhance the quality of life of Londoners. Recognizes that rising temperatures may reduce winter deaths and that warmer summers may benefit tourism, which is an important component of the London economy.	Nickson *et al.* (2011)
Massachusetts, United States	This report provides an overview of the observed and predicted changes to Massachusetts' climate key adaptation strategies that cut across multiple sectors. The report also describes the vulnerabilities to climate change and the adaptation strategies that could help five broad areas, including components of the local economy such as tourism, increase resilience and preparedness.	Executive Office (2011)

3.5 Adaptation policy making and planning at multi-country and international levels

There are few multi-country and international policy and planning initiatives that deal with managing the risks climate change poses to tourism (OECD and UNEP, 2011). Those of greatest relevance are described in the following three case studies or as the additional examples.

Case study: Increasing the weather and climate resilience of rail infrastructure and operations

To assist operators to respond appropriately to climate change the International Union of Railways is implementing a project, Adaptation of Railway Infrastructure to Climate Change (ARICC). It focuses on the integrated management of the weather- and climate-related natural hazards. Several members of the International Union with experience of extreme weather events and conditions have joined forces to gather existing knowledge, based on these experiences, in order to develop new understanding and management capabilities related to the impact of climate change on rail infrastructure and operations. The project starts from a base of managing today's weather conditions and develops solutions and strategies to prepare for future weather/climate conditions (Figure 3.3). The aim is to maintain and improve railway infrastructure and operational performances by avoiding and minimizing damage to railway infrastructure.

The study is providing information on risk assessments, potential damages and the costs of adaptation. Good practices have been documented, as have options for adopting standards to climate-proof new and existing rail infrastructure. In addition, a toolbox that can be used to identify vulnerable assets and locations has been developed.

Case study: European adaptation and tourism frameworks

The White Paper 'Adapting to Climate Change: Towards a European Framework for Action' (or 'Adaptation Framework') was released in 2009 (Commision of the European Communities, 2009). It presents the framework for adaptation measures and policies to reduce the EU's vulnerability to the impacts of climate change. The Framework will complement action by Member States and support wider international efforts to adapt to climate change, particularly in developing countries. The objective of the Adaptation Framework is to improve the EU's resilience to and ability to deal with the impact of climate change.

The following year, the European Commission released the Framework for Action for Tourism in Europe, or 'Tourism Framework' (Commision of the European Communities, 2010). Significantly, the Tourism Framework makes no reference to the Adaptation Framework, despite identifying that the supply of tourism services must take into account constraints such as those linked to climate change, the scarcity of water resources and pressure on biodiversity. It thus

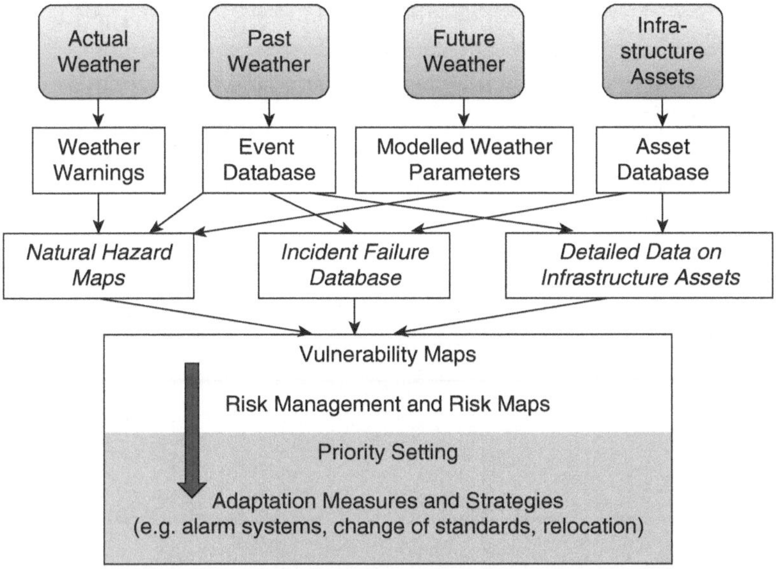

FIGURE 3.3 The approach used in the ARICC Project to identify priority adaptation strategies and measures to increase the weather and climate resilience of rail infrastructure and operations. (After Nolte, 2010)

recognizes that tourism businesses need to reduce their use of drinking water where there is a risk of drought, and reduce their greenhouse gas emissions and environmental footprint.

The Tourism Framework acknowledges that tourism's competitiveness is closely linked to its sustainability, as the quality of tourist destinations is strongly influenced by their natural and cultural environment and their integration into a local community. Thus one of the four priorities in the Tourism Framework is to promote the development of sustainable, responsible and high-quality tourism. A specific action related to this priority is to facilitate identification by the European tourism industry of risks linked to climate change in order to avoid loss-making investments, and explore opportunities for developing and supplying alternative tourism services.

Case study: Managing the effects of climate change on world heritage

The Strategy to Assist States Parties to Implement Appropriate Responses to Manage the Effects of Climate Change on World Heritage (World Heritage Convention Secretariat, 2008) calls for the management plans of all sites potentially threatened by climate change to be updated to ensure the sustainable conservation of their outstanding universal values is not compromised. The plans should be based on appropriate monitoring and vulnerability assessments. Tailored risk reduction measures should also be undertaken as required. Lessons learned at several sites worldwide show the relevance of designing and implementing appropriate adaptation measures. These might include increasing the resilience of a site by reducing non-climatic sources of stress, modifying boundaries and buffer zones to facilitate migration of species, preventively draining a glacial lake to avoid the occurrence of an outburst flood, improving dykes to prevent coastal flooding, and supporting traditional methods to protect a site from sand encroachment. A well-known example of a World Heritage Site affected by climate change is Mount Everest (Tibet/Nepal) (Plate 3.6).

PLATE 3.6 Sagarmatha National Park (Mt Everest) seen from the Tibet side. Increased melting potentially affects climbing and other tourist operations. (S. Becken)

Other examples of multi-country and international tourism adaptation policy and planning initiatives

The examples opposite describe some of the limited number of multi-country and international policy and planning initiatives that deal with managing the consequences and opportunities climate change presents for tourism.

3.6 Conclusion

Planning for climate change, rather than reacting to it, ensures that more response options are available, and the costs of adaptation are often reduced. But preparing adaptation policies and plans means addressing the uncertainty as to how the climate will change in the future, and how tourism and other systems will respond. The nature of changes in the climate to which tourism operators, tourists and other

Policy	*Description*	*Source*
Multi-Hazard Contingency Planning Manual for the Caribbean Tourism Sector	The manual is designed to help the industry to be better prepared for dealing with critical threats to tourism such as hurricanes and terrorist incidents.	www.onecaribbean.org/content/files/OASDisasterManual2009final.pdf
Framework for Regional Destination Adaptation to Climate Change	The framework is at the local or regional destination level, and not only takes a risk management approach to adaptation, but also recognises that climate change may present opportunities for future development; emphasis is on the role of the consumer, in recognition of the comparatively high adaptive capacity of tourists.	Jopp *et al.* (2010)
Davos Declaration 2007	Agreed in 2007, it urges action by the entire tourism sector to face climate change, considered to be one of the greatest challenges to sustainable development, and to achieving the Millennium Development Goals in the twenty-first century.	www.unwto.org/pdf/pr071046.pdf
World Travel & Tourism Council	Provides recommendations for actions needed at national and local levels to address climate change; these could be incorporated into existing national frameworks.	www.wttc.org/bin/pdf/original_pdf_file/climate_change_-_a_joint_appro.pdf
Decision 1/CP.16 of the Conference of the Parties to the UNFCCC	Strategic Element 2 seeks to strengthen the climate resilience of the most vulnerable sectors (tourism, water supply, coastal and marine ecosystems, coastal infrastructure, health) by supporting the dissemination of successful adaptation measures.	www.cpdcngo.org/cpdc/attachments/article/107/Regional%20Framework.pdf
Framework for Climate Change Adaptation in the Tourism Sector	The framework represents an integration of relevant best practice and the commonly recommended components of frameworks that have been applied in several developing nations and in several economic sectors, though none has focused explicitly on the tourism sector.	Simpson *et al.* (2008b)

stakeholders must respond is strongly dependent on the emission and sequestration rates of GHGs. Past and current emissions are known, but future emissions will be determined by social and economic development choices. In addition, policies and plans that focus on adaptation have to consider complex interrelationships, including dynamic physical and ecological responses in both the more immediate and longer terms, as well as the uncertain responses by tourists themselves.

Policy making and planning how the tourism sector will adapt to climate change is typically coordinated and led by national governments. Since tourism is an activity that involves many sectors, directly and indirectly, there must be strong coordination in order that the linkages are addressed in ways that ensure efficient responses and long-lasting results. Institutional structures which do not facilitate a comprehensive overview of the tourism sector will impose their own limitations on planning and implementation, not only to the detriment of tourism itself, but also to the many benefits it could bring to the economy and society. Thus governments have a role to play in ensuring that strong institutions are in place and that they are capable of delivering effective coordination of the tourism system across the public and private sectors, including civil society.

Large tourism corporations and tourism industry organizations also have key roles in ensuring that adaptation policies, plans, decisions and implementation are well informed and coordinated. Smaller tourism operators, and tourists themselves, should benefit from these higher-level policies, plans and institutional arrangements. Smaller operators might well prepare an adaptation plan as part of their wider business plan, especially if higher-level policies and plans indicate their business is at significant risk. Examples of such initiatives will be presented and discussed in the following chapter (Chapter 4).

In most countries preparation of the policies and plans that describe how the tourism sector will manage the risks posed by climate change is still in its infancy. There is considerable opportunity for governments, the private sector, and civil society to be more fully involved. For many tourism-dependent countries the window of opportunity is already closing. Climate variability and change, including extreme events, are already having adverse impacts, while the economic and financial opportunities are being missed. Governments and the tourism industry can learn much from each other, as well as from other sectors that are more advanced with their adaptation policies and plans.

The current low awareness of how the components of the tourism sector might best adapt to climate change, the lack of knowledge regarding the complexities of tourism–climate interrelationships, and the challenges to implementing effective whole-of-government and whole-of-country responses to addressing climate risks to the tourism sector, highlight the need for national tourism adaptation policies and plans. These should aim to improve understanding on how the tourism and related sectors will be affected by climate change, and how adaptive capacities will be enhanced through such activities as targeted research, education, training and institutional strengthening at national, sub-national and industry levels.

4

ADAPTATION PRACTICE

4.1 Introduction

The climate is changing. Chapter 2 shows that it is changing faster, and in more and different ways, than had been anticipated. In addition to trying to slow the rate of future changes (see Chapters 5 and 6), we have two response options available to us (Figure 4.1). These are both referred to as 'adaptation', a term used to describe actions that aim to moderate, cope with and benefit from the consequences of climate change in order to manage risk and reduce vulnerability (Jopp *et al.*, 2010).

We can be forced to adapt. This occurs when the prevailing consequences of climate change are found to be unacceptable. As we shall see, some tourism operators and tourists are already experiencing this situation. On the other hand, we can act in a proactive manner, by anticipating the changes. This involves planning and then implementing actions designed to either reduce the risks posed by climate change or exploit the opportunities. Illustrative examples will also be presented in the following sections.

Even though considerable effort has been spent conceptualizing and piloting adaptation, the majority of studies have been limited to vulnerability and risk assessments, with emphasis on natural systems and planning, as opposed to tangible actions to reduce climate-related risks. Arnell (2010) assessed the current levels of understanding and practical experience of adaptation by reviewing all adaptation related articles in the journal *Climatic Change*. He concluded that there are very few published examples of how adaptation to climate change is actually being delivered, and also few articles on the barriers that will influence how adaptation takes place. Based on a more comprehensive assessment, Berrang-Ford *et al.* (2011) concluded that, while considerable research on adaptation has been undertaken, the majority of publications describe vulnerability assessments, natural systems and intentions to act, not tangible adaptation actions. Adaptation action is more frequently reported in

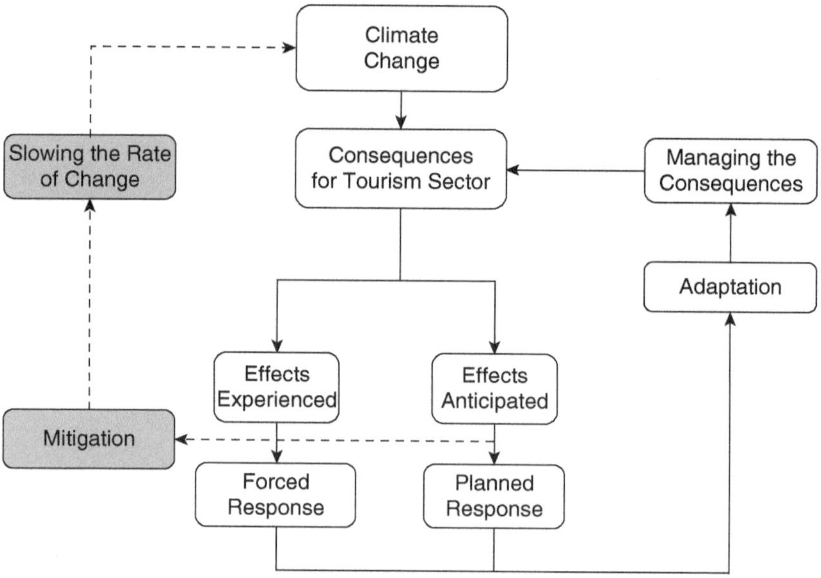

FIGURE 4.1 Forced and planned adaptation – the two categories of actions designed to reduce the consequences of climate change. Also shown for completeness is mitigation, the third response to climate change.

developed nations, with middle-income countries underrepresented and low-income regions dominated by reports from a small number of countries. Proactive adaptation is the most common adaptive response, particularly in developed nations. While similar assessments specific to adaptation in the tourism sector have yet to be undertaken, the examples of adaptation practice presented in this chapter will tend to support the wider conclusions of Berrang-Ford *et al.*

There are many reasons specific to the tourism sector that explain the need for action to address both the current and anticipated consequences of climate change. Of course the sector is not the only one being impacted. Nor does it respond alone. But this chapter adopts a narrower lens, to allow a more practical focus. It responds to the observation of Simpson *et al.* (2008b) that the tourism sector has been overlooked in efforts to identify adaptation options. As recently as 2010, Scott and Becken noted that climate adaptation research in the tourism-recreation sector is five to seven years behind that of sectors which are being more proactive.

Figure 1.3 in the Introduction (Chapter 1), frames the following discussion on how the tourism sector is already adapting to climate change, and the further actions that can be taken to reduce impacts and exploit opportunities. Studies have shown that the main consequences of climate change for the tourism sector, and where there is a need and opportunities to act, relate to destinations becoming more or less appealing to tourists, impacts on transport infrastructure and operations, threats to tourism's resource base that include natural landscapes and ecosystems, changed risks and experiences while at a destination, and the sustainability of tourism enterprises and destinations (Becken and Hay, 2007).

> There is growing evidence since the IPCC Third Assessment of human activity to adapt to observed and anticipated climate change. … Even if atmospheric greenhouse gas concentrations remain at 2000 levels … there are some impacts for which adaptation is the only available and appropriate response. (IPCC, 2007a)

4.2 Incentive to travel, and appeal of potential destinations

Anticipated consequences of climate change

Climate change alters the comparative advantages of holiday destinations, creating opportunities for some, and challenges for others. The appeal of a destination to tourists depends, in part, on the attributes of competing destinations. UNWTO *et al.* (2008) provide a summary assessment of the most at-risk tourism destinations for the mid- to late twenty-first century, including the Caribbean, the Mediterranean, and the Indian and Pacific Ocean islands. However, the very limited information available on the potential impacts of climate change for some tourism regions means that qualitative assessments such as this must be considered with caution.

Arguably some of the most dramatic changes in destination appeal are exemplified by so-called 'last chance tourism' where tourists explicitly seek vanishing landscapes or seascapes, and/or disappearing natural and/or social heritage (Lemelin *et al.*, 2010). These include destinations already impacted, or projected to be impacted, by the adverse effects of climate change (Huebner, 2011). The increased number of tourists attracted to these destinations may in fact hasten their demise, a notion that will be explored further in Sections 4.4 and 4.6. Examples of 'last chance tourism destinations' include those in the polar regions (focusing on the notion of disappearing polar bear habitat in the Arctic and glaciers and ice shelves in the Antarctic) and low-lying island nations in the Pacific and Indian Oceans. The vulnerability of these latter destinations to climate change has been highlighted in terms of sea-level rise, coastal erosion, water and food scarcity, flooding, and coral bleaching, leading to descriptions such as 'titanic states' (the sinking or drowning islands), 'high tide', 'dark clouds over paradise' and 'canaries in the coalmine' (early warnings) (Barnett and Campbell, 2010).

> *Sun, Sea, Sand, Sinking?*
>
> Software giant Microsoft has come up with a new angle to encourage people to visit islands like ours – some won't be around much longer.

Helenka Bednar, a contributor to MSN Travel, part of the Internet portal formerly known as The Microsoft Network, writes: 'The sea has gobbled up two islands [between India and Sri Lanka] in the past few months alone – make sure you visit some of the world's most idyllic spots before it's too late.' Bednar says many more islands are predicted to disappear over the next few decades due to global warming. 'If you blink, you won't miss them, but we'd advise paying a visit to these floating places of paradise while you still can. From the turquoise glow of the Maldives to the melting ice of Eskimo village Shishmaref, island-hop with our pick of scenic places headed for a watery exit.' The section dedicated to the Cook Islands is titled 'Ticking clock: wave surges are threatening this haven'. 'Palm trees, warm seas and bright azure lagoons – the Cook Islands are our idea of paradise. But storms, flooding, wave surges and rising sea levels are not working in their favour. The islands' coral reefs provide a natural defence against storms and flooding too, but industrial growth is threatening their existence.' (Cook Islands News, November 3, 2011)

Of all the players in the tourism system, tourists themselves have the greatest ability to respond to climate change impacts, by substituting the place, timing and type of holiday (Jopp *et al.*, 2010). Studies using global tourism simulations models (e.g. Bigano *et al.*, 2008) show an overall response to climate change, with patterns of tourism shifting towards higher altitudes and latitudes. These models are now being downscaled to national and sub-national resolutions. For example, Hamilton and Tol (2007) found that, because of climate change, tourists from Germany, Ireland and the United Kingdom will spend more holiday time in their own country. Within Ireland and the UK the pattern of shifting demand is similar to that at the international level, namely that tourism shifts pole-ward. However, within Germany the opposite change is observed – tourism shifts south, due to the continental interior warming faster than coastal areas.

But the findings of such high-level analyses need to be treated with caution. For example, Moreno (2010) reports that, while the Belgian and Dutch attach great importance to climate as a destination attribute, for beach tourism specifically the absence of rain is more important than a 'comfortable temperature'. Significantly, heatwaves were considered the least important climate change impact, being seen as 'not too negative'. Only a few respondents considered high temperatures as unfavourable. Even if climate change results in the ideal weather for beach tourism (temperatures of approximately 28°C, light breeze, and a blue sky) occurring in northern Europe, the large majority (72 per cent) of respondents said they would still travel to the Mediterranean for their beach holidays. Meanwhile other studies, such as that undertaken by Rosselló-Nadal *et al.* (2011) for the United Kingdom, show how current and anticipated weather and climate conditions where the tourists live affect outbound tourist flows.

The realities, as opposed to the potential changes described above, will depend to a large extent on the adaptability of both institutions and tourists as they anticipate and respond to climate change. The actual changes could be substantial in terms of both the spatial and temporal redistribution of tourism activities the world over. By way of case studies, possible adaptation interventions will be described in relative detail for hotter summers in the Mediterranean, including significantly more heatwaves, and for less snow cover and duration in mid-latitude ski areas. Less detail will be provided on other actual or planned adaptation initiatives to manage the influence of climate change on incentive to travel and on the appeal of potential destinations.

Case study: Hotter summers and more heatwaves in the Mediterranean

Climate change projections suggest that European summer heatwaves will become more frequent and severe during this century, consistent with observed trends in the past decades (Fischer and Schär, 2010). High-resolution regional climate simulations using six climate models showed that the average number of heatwaves in Europe is projected to increase from one per three to five summers in 1961–1990, to two or three every summer in 2071–2100 (Plate 4.1).

Consistent with the fact that the Mediterranean basin is one of the most sensitive areas of the world to human-induced climate change (Calbó, 2010), it is not surprising that southern European river basins and the Mediterranean coast are projected to be most affected by changes in heatwave frequency and duration. More detailed modelling reveals that the frequency of heatwave days for the Iberian Peninsula and Mediterranean region is expected to increase from about two days per summer for 1961–1990, to around 13 days by 2021–2050, and to as many as 40 days by 2071–2100. The most densely populated European regions, including prime tourist destinations such as the urban areas of Athens, Naples, Rome, Milan and Bucharest, are projected to experience the greatest increases in health risk.

Given such projections, Rutty and Scott (2010) examined the possibility that during the peak summer season the Mediterranean region conditions will become uncomfortably hot for tourists. Perceptions of 'unacceptably hot' for comfortable tourism activities at beach and urban destinations were quantified for the young adult travellers by way of a survey of 850 university students in five countries that represent source markets for the Mediterranean. The threshold that defined unacceptable heat for the majority of respondents was then compared against thermal conditions (temperature and humidity) in a baseline climate (1961–1990), and for early (2011–2035), mid- (2046–2065) and late (2080–2099) century climate scenarios for ten Mediterranean destinations. Even under the warmest early century climate change scenario no additional beach or urban destination was 'unacceptably hot'. By mid-century, thermal conditions for a further two beach destinations and

PLATE 4.1 Climate change is expected to result in intolerably hot conditions for mid-summer sunbathers, such as these on the coast of Croatia. (H. Henry)

one urban destination became 'unacceptably hot' during the peak summer months, while by late century four beach and five urban destinations exceeded the heat thresholds in the summer months.

Rutty and Scott make two important observations based on these findings. First, the fact that most Mediterranean destinations will not become 'unacceptably hot' until towards the end of the twenty-first century provides ample time for northern European travellers to acclimatize to warmer average temperatures at home. Thus the temperature associated with the 'unacceptably hot' threshold may increase with time, at least partially offsetting the increase in maximum summer temperatures in the coming decades. Second, in the coming decades there will be a decrease in the number of months considered to be 'unacceptably cool' for beach and urban vacations, while the number of months considered to be 'ideal' will increase. These two points highlight, in turn, two longer-term adaptation options among others that are shown in Figure 4.2. One involves a change in behaviour, with tourists from northern Europe deciding to take winter and shoulder-period, rather than summer, vacations at Mediterranean destinations. On the other hand, some tourists will adapt autonomously to the changing climates, and will thus maintain their traditional holiday travel patterns, despite higher temperatures at their destinations.

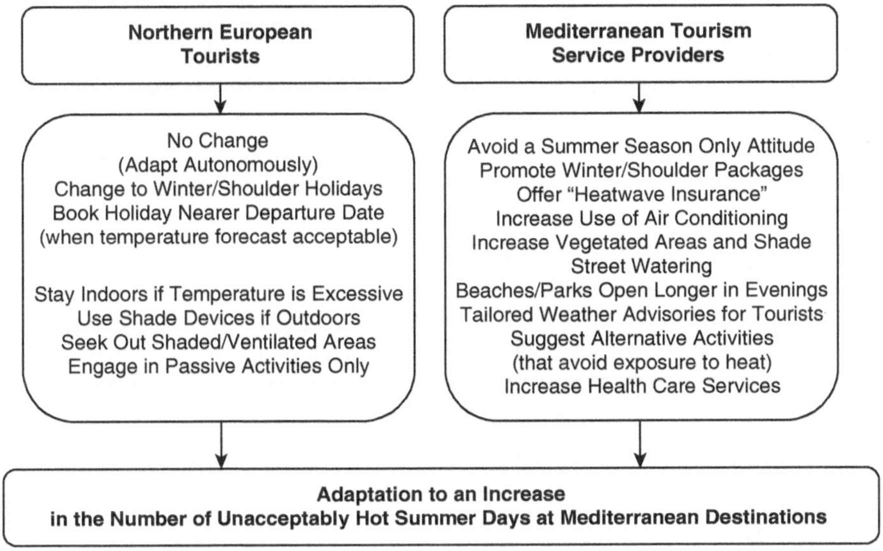

FIGURE 4.2 Adaptation initiatives northern European tourists and Mediterranean tourism service providers might take as a result of an increase in the number of unacceptably hot summer days at Mediterranean destinations.

Case study: Less snow cover and duration in mid-latitude ski destinations

Mountain and winter sports, and especially skiing, are frequently identified as being highly vulnerable to climate change, including by UNWTO *et al.* (2008). For example, Hoy *et al.* (2011) report that favourable conditions for winter tourism have declined over recent decades in the Saxon low mountain ranges of Germany. This, and projections of continuing warming, is necessitating a change to a wider variety of winter tourism activities from those that are predominantly snow based. Beniston *et al.* (2011) show that, while the Swiss Alps will not be totally devoid of snow-abundant winters in the foreseeable future, they may occur less than 5 per cent of the time in the latter 30 years of this century. Scott *et al.* (2011) report similar findings for the Northeast United States, with serious economic implications for businesses and communities. Under one climate change scenario, reliable snowmobile seasons (at least 50 days of snowmobiling possible) will be rare by 2070–2099.

In terms of consequences for the industry, Scott and McBoyle (2007) concluded that, since most assessments do not consider the adaptive capacity of the ski industry, it is difficult to determine the degree to which the multi-billion-dollar winter tourism industry is at risk from climate change. Only four of the 15 studies reviewed by Scott and McBoyle took into account adaptation measures that could be implemented by ski area operators. As a result, many of the recent estimates of future damages may be overestimated.

As the following examples show, many adaptation measures have the potential to significantly reduce the negative impacts of climate change on the ski industry. But as the examples also show, there are limits on the ability of the ski industry to adapt, due in part to economic and resource constraints.

The already large investment in snowmaking in the Northeast United States has substantially reduced the vulnerability of the ski industry, with climate change posing a high level of economic risk to only four of the 14 ski areas in 2010–2039 (Scott *et al.*, 2011). However, by 2070–2099 all but four of these ski areas will suffer a high level of economic risk. As snowmaking increases, issues such as water availability and cost will become more significant. Pickering and Buckley (2010) highlight similar issues for the multi-billion-dollar Australian ski industry, where studies have suggested that by 2020 natural snow cover will become inadequate at 65 per cent of the country's ski resorts, leading to plans for additional snowmaking. It is anticipated that over 700 additional snow guns will be required by 2020, representing a capital investment of around US$100 million, use of 2,500 to 3,300 million litres of water per month, and increased energy consumption. Pickering and Buckley argue that, since this is not practically feasible, especially given increasing water constraints, low-altitude ski resorts may not be able to rely on snowmaking even for adaptation to climate change in the short term. Rather, they are likely to convert to summer activities and land sales.

However, Bark *et al.* (2010) report that Snowbowl, a ski resort in Arizona, has overcome such water constraints by contracting with a municipality to use treated wastewater. As a result, even when taking into account temperature projections relative to technical thresholds for the manufacture of snow, snowmaking will likely remain feasible out to 2030, a period coincident with the typical lifetime of ski-field investments. But by 2050, temperatures will likely exceed technical thresholds for snowmaking, at least in the shoulder seasons. Thus in years when natural snowfalls are poor, the ski season may be curtailed. By 2080, without snowmaking efficiency improvements, warmer temperatures will make snowmaking increasingly more expensive. As a result, resort managers will need to plan for a longer-term future where operations and snowmaking are shifted to higher elevations on areas are shaded.

In a study to explore potential adaptation options to climate change impacts in low mountain areas in Eastern Germany, Hoy *et al.* (2011) found that, while diversifying tourism options presents opportunities for businesses, significant obstacles do exist. For instance, although the majority of tourism operators interviewed in the study recognized climate change to be a serious issue, almost the same number of interviewees were not aware of the extent to which snowfall is projected to change in the region over the next 15–20 years. Uncertainty about the nature of projected changes also served as a barrier to action. None of the alternative activities that downhill ski operators could adopt in response to declining snowfall are seen as profitable. Also, they are often perceived by younger tourists as being less exciting than downhill skiing.

However, as an increasing number of older tourists are visiting the region, there is an opportunity for tourism operators to work on attracting groups who are interested in the beauty of mountain landscapes regardless of wintry conditions. Marketing of less snow-intensive activities should be done very positively, so they are perceived as valuable activities in their own right, and not seen as substitutes for deteriorating conditions.

Other examples of adaptation to changing destination attractiveness

The following list provides some additional examples of adaptation initiatives undertaken in relation to managing the consequences of climate change with respect to incentive to travel, as well as the appeal of potential destinations.

Event/threat	*Adaptation example*	*Source*
Hurricanes disrupting tourist arrivals and convention business, United States	The Florida Recovery Office was established after eight named storms made landfall in Florida during the 2004 and 2005 hurricane season. The State allocated US$30 million to 'hurricane recovery' marketing and developed a weather insurance programme for convention organizers, whereby the State pays the premiums for US$200,000 of insurance to cover rescheduling costs associated with hurricane disruption.	UNWTO *et al.* (2008)
BaltCICA, Germany	Climate Change: Costs, Impacts and Adaptation in the Baltic Sea Region: The BaltCICA project, involving local and regional partners, prepares regions and municipalities to cope with a changing climate.	www.baltcica.org/
RADOST, Baltic Coast, Europe	Regional Adaptation Strategies for the German Baltic Coast: the implementation project 'Tourism in times of climate change – Regional adaptation strategies' is focused on the Rostock area. It develops adaptation strategies for beach-based tourism, the region's main attraction.	http://klimzug-radost.de/en/project/info/implementation-projects/tourismus-and-climate-change-regional-adaptation-strategies
ClimAlpTour, Europe	The ClimAlpTour project aims at providing a novel, science-based approach to tackle the impacts of climate change on Alpine tourism.	www.climalptour.eu/content/

4.3 Transport infrastructure and services used by tourists

Anticipated consequences of climate change

Tourism's very existence depends on transport, whether this is in terms of moving people and tourism supplies from their origin to their destination, or as an integral part of the tourism experience at the destination. The ability of the transport system to deliver those services depends, in part, on weather and climatic conditions. Transportation infrastructure is designed for typical weather patterns, reflecting not only the local climate but also incorporating assumptions about a reasonable range of temperatures, precipitation and other environmental conditions.

Climate change is now one of the many concerns that must be addressed in planning for improvements in transportation and the built infrastructure. The effects of climate change will manifest themselves in many ways, including through increases in weather and climate extremes and rising sea levels, coupled with storm surges and land inundation. Transport infrastructure and operations will be affected most by those climate changes that cause environmental conditions to extend outside the range for which the system was designed.

The consequences will vary by mode of transportation and location, but they will be widespread. They could also be costly, in both human and economic terms, unless there are significant and timely changes in the planning, design, construction, operation and maintenance of transportation systems. Illustrative consequences of relevant changes in climate are provided in Table 4.1. URS (2010) provides additional examples of the impacts of climate change on transportation.

The consequences described in Table 4.1 highlight the interdependencies in infrastructure. These often lead to 'cascade failure', where the failure of one aspect of infrastructure, such as flood defences, leads to other failures. For example, subsequent flooding of a power station might result in power cuts that affect telecommunications systems. It is therefore important to manage the consequences of climate change for transport infrastructure and operations in a comprehensive and systematic manner, especially as these systems are becoming increasingly interdependent (Royal Academy of Engineering, 2011).

Case study: Enhancing the resilience of built infrastructure to climate change

To make decisions today about rehabilitating or retrofitting transportation facilities, especially those with long design lives, transportation planners and engineers must consider how the anticipated changes in climate will affect these facilities 50 years or more from now. Adapting to climate change also requires re-evaluation, development and regular updating of standards that guide infrastructure design (Transportation Research Board, 2008). ADB (2010) distinguishes adaptation options in the transportation sector as engineering (structural) options (subsurface conditions, material specifications, cross-section and standard dimensions, drainage

TABLE 4.1 Changes in climate and their possible consequences for transportation. (After Transportation Research Board, 2008)

Change in climate	*Consequences for operations*	*Consequences for infrastructure*
Increases in very hot days and in heatwaves	Impact on lift-off load limits at high-altitude or hot-weather airports with insufficient runway lengths, resulting in flight cancellations or limits on payload.	Thermal expansion on bridge expansion joints and paved surfaces. Concerns regarding pavement integrity (e.g. softening), traffic-related rutting, migration of liquid asphalt. Rail-track deformities.
Rising sea levels, combined with storm surges	More frequent interruptions to coastal and low-lying roadway travel and rail services due to storm surges; More severe storm surges, requiring evacuations, or changes in development plans; Potential for closure or restrictions at several major airports that lie in coastal areas, affecting service to some areas with high populations.	Inundation of roads, rail lines and airport runways in coastal areas. More frequent or severe flooding of tunnels and infrastructure in low-lying areas. Erosion of road base and bridge supports. Reduced clearance under bridges. Changes in harbour and port facilities to accommodate higher tides and storm surges.
Increases in intense precipitation events	Increases in weather-related delays, and in traffic disruptions; Increased flooding of evacuation routes; Increases in airline delays due to convective weather.	Increases in flooding of roadways, rail lines, tunnels and runways. Increases in road washout, damage to rail-bed support structures, and landslides and mudslides that damage roadways and rail tracks.
More frequent strong cyclones (Categories 4 and 5)	More frequent interruptions in air service; More frequent and potentially more extensive emergency evacuations; More debris on road and rail lines, interrupting travel and shipping.	Greater probability of infrastructure failures. Increased threats to stability of bridge decks. Impacts on harbour infrastructure from wave damage and storm surges.
Increases in Arctic temperatures	Longer ocean transport season and more ice-free ports in northern regions; Possible availability of a northern sea route or North-west passage for cruise ships.	Thawing of permafrost, causing subsidence of roads, rail beds, bridge supports, pipelines and runway foundations. Shorter season for ice roads.

and erosion, and protective engineering structures), and non-engineering options (maintenance planning and early warning, alignment, master-planning and land-use planning, and environmental management). In addition, in some circumstances, not responding to climate change, or 'do-nothing', might also be an option. This could include, for example, allowing infrastructure to deteriorate and eventually be decommissioned, instead of working to climate-proof the infrastructure.

There are many options for increasing the resilience of transportation infrastructure to climate change. These have been identified in numerous studies such as the Transportation Research Board (2008) and the IPCC (2007a). Adaptation interventions cover both design and operational changes, and include increases in the standard for drainage capacity for new transportation infrastructure and major rehabilitation projects, improvements in the monitoring of conditions and issuance of real-time messages to users such as motorists, removal of traffic bottlenecks on critical evacuation routes, building of more system redundancy, adoption of modular construction techniques where infrastructure is in danger of failure and greater use of heat-tolerant street and highway landscaping.

Case study: Railway maintenance and operations

The threat of natural hazards has always been an integral factor in the planning, construction and operation of railways. However, much long-distance rail infrastructure was constructed more than 100 years ago. Moreover, in many places the rail tracks and other vital infrastructure have suffered from a lack of appropriate maintenance, largely due to financial cutbacks. Given that the frequency and intensity of incidents caused by extreme weather events are expected to increase in the future, there will be increases in both the risks to rail systems achieving performance targets and in the costs of maintaining performance. If appropriate preventive measures are taken at the appropriate time, the risks and costs will be manageable.

The main effects of climate change on railway operations in the UK in the next few decades are likely to be an increase in track buckling, severe strain on railway drainage systems and the increased likelihood of disruption because of extreme weather events (see Table 4.2 for the case of high temperatures) (Rail Safety and Standards Board, 2010; Baker *et al.*, 2010; Rossetti, 2002).

Heatwaves, flooding and high winds are already having detrimental effects on operating costs and performance. The threat of climate change to the rail system is receiving increasing attention. For example, the United Kingdom's Climate Change Act 2008 (Government of the United Kingdom, 2008) requires the Office of Rail Regulation and Network Rail to assess and report on the risks of climate change to rail operations, and to formulate plans to manage the risks. Network Rail's work on adaptation has highlighted the importance of prediction, not only to facilitate required long-term management decisions, but also to support the reliable operation of the railway through the maintenance, modification or replacement of potentially vulnerable assets. Enhancing the resilience of railway assets to current extreme

TABLE 4.2 Asset and operational risks to Network Rail due to heat, including exacerbation by climate change. (After Network Rail, 2010)

Risks associated with heat	*Safety impact*	*Performance impact*	*Long or short term?*
Air conditioning failure in carriages	Low	Medium	Short
Track buckling	High	High	Long
Speed restrictions due to buckle risk	Medium	High	Long
Use of heat watchmen for buckle risk	Low	High	Long
Floating electrical earth caused by a low water level	High	High	Long
Reduced window of opportunity for work – renewal and maintenance – due to heat restrictions on track	Low	Medium	Long
Reduction in track quality due to less maintenance	Low	Medium	Long
Staff working conditions in hot weather	High	High	Long
Earthworks desiccation	Low	Medium	Long
Effect of heat on swing bridges	Low	Low	Long
Contact wire sagging at terminal stations	Low	High	Long
Reduced transformer life	Low	Low	Long
Solar gain affecting line-side equipment – signalling, power, telecoms	Medium	High	Short

weather events, and to future climatic conditions, requires the assessment of the railway as a whole system (Network Rail, 2010).

The United Kingdom Meteorological Office is providing forecasts of adverse weather conditions that can cause chaos for train operating and maintenance companies as a result of collisions, derailments, cancellation of services and overheating. Operational forecasts include site-specific predictions for railways and stations. These enable operators to make timely decisions to improve planning and resources for weather-related events which could affect the safety and punctuality of services. Forecasts of leaf fall provide site-specific data where rail adhesion may be compromised (Plate 4.2). The weather elements featured include precipitation, wind and leaf-fall risk.

Overtopping can make rail lines that run close to the sea impassable, causing train derailments in severe cases. Overtopping forecasts provide details of prevailing sea conditions (including wind, water level, swell direction, wave height and wave period) both near-shore and inshore at sea defences, providing a warning of the risk of waves encroaching on the track. Rail ice forecasts provide information on the risk of icing and the risk of freezing rain, sleet and rail temperature, for a given location, including a daily view of hazards and an outlook up to five days ahead. Critical thresholds can be set by operators for variables such as high temperatures, snow, gales or frost. Warnings are issued when these thresholds are at risk of being exceeded (Met Office, 2010).

PLATE 4.2 Rail line subject to high levels of leaf fall and resulting low adhesion. (J. E. Hay)

Case study: Adaptation of flight operations

A change in climate will modify the demand for air transport, increase the threats to aviation infrastructure and bring changes to air transport operations (EUROCONTROL, 2008). Figure 4.3 identifies the weather and related phenomena that can be influential for given phases of flight operations. Overall, weather is becoming a dominant cause of flight delays in Europe. In 2005 weather accounted for 27 per cent of total delays, and for 45 per cent of airport delays (EUROCONTROL, 2006). On hot summer days, at high-altitude airports, it may be necessary to restrict takeoff weights or even cancel flights. Airport runway designs take into account the usual conditions of temperature, moisture, wind velocity and visibility. This means that, in general, higher-altitude airports have longer runways. However, with an increase in the number of days with higher temperatures, the incidence of limited operations at high-altitude airports will increase (Peterson *et al.*, 2008). On the other hand, many airports in higher-latitude locations will likely benefit from reductions in the cost of snow and ice removal and associated delays, and in the environmental impacts of salt and chemical use. Airlines will also benefit from reductions in the need for de-icing of aeroplanes. Climate projections suggest that there will be a northward shift in the storm tracks, making northern European airports potentially more vulnerable to severe weather closures and delays.

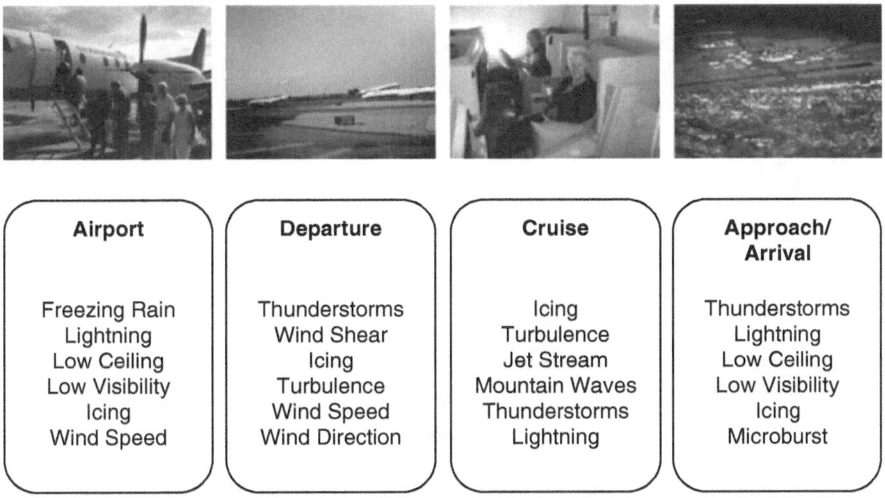

FIGURE 4.3 Weather and related phenomena that can be influential for given phases of flight operations. (After Pejovic, 2008)

Pejovic *et al.* (2009) have assessed the impact of climate change on air transportation operations, using London's Heathrow Airport (England) as a case study. Analysis of weather-caused delays at the airport for the period January 2001 to May 2006 revealed that the most frequent reported cause of delay was due to airport and/or runway weather-related restrictions at the destination airport (Figure 4.4). Airport-related delays represent 75 per cent of the total reported delays or 91 per cent of the total delayed minutes. The remaining delays related to en-route conditions. Thus weather impact is strongest during takeoff and landing.

Thunderstorms, snow and fog increased the chance of weather-related delays by more than 25 per cent, while a one knot increase in wind speed above the mean increased the probability of delay by 8 per cent. Such statistical relationships were used in conjunction with seven climate models to estimate weather-related delays in the 2050s. By that time the likelihood of experiencing a delay due to adverse weather conditions is increased by up to 7 per cent during the winter months. However, during the summer months the frequency of delays is expected to decrease by almost 15 per cent. These findings are relatively insensitive across the three greenhouse gas emission scenarios used in the study.

WeatherBill Inc. (2008) analysed the contribution of weather conditions to flight disruptions for major airlines and large airports in the United States. Irrespective of season, location or airline, precipitation was found to have a stronger effect than temperature. Unusually warm temperatures are associated with reduced delays in winter and increased delays in summer. In summer, the United States frequently experiences substantial flight disruptions as a result of severe convective

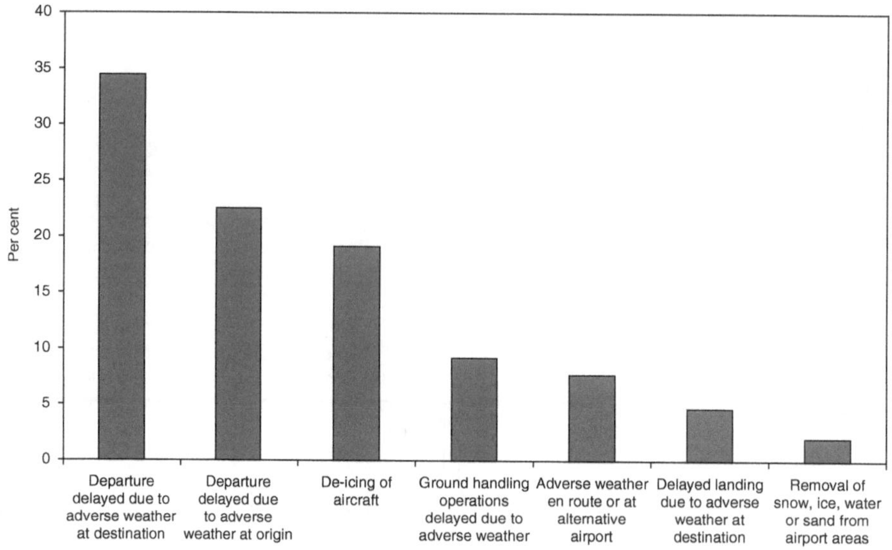

FIGURE 4.4 Causes of weather-related delays at Heathrow Airport, for the period January, 2001 to May, 2006. (After Pejovic, 2008)

weather that obstructs en-route and terminal airspace (Klein *et al.*, 2009). Flight cancellations, diversions and excess miles flown increase the direct operating costs for air carriers by billions of dollars. Klein *et al.* estimated the avoidable portion of arrival delays and cancellations due to terminal weather forecast inaccuracy and found that total benefits of improving terminal forecast accuracy would be at least US$330 million per year. This represents a highly cost-effective adaptation option.

In the United States a large proportion of flight delays and cancellations are due to thunderstorms and other convective weather systems. Their likely increasing impact on commercial aircraft operations is one reason for the considerable effort now underway in the United States to increase airspace capacity and flexibility using enhanced air traffic management (ATM) systems. Traffic flow managers and commercial airline dispatchers are now using the recently developed Future ATM Concepts Evaluation Tool (FACET) for real-time operations planning (Bilimoria *et al.*, 2000; NASA, 2010). FACET integrates live air traffic and weather data, allowing operators to determine wind-optimal routes, and reroute flights around congested airspace and severe weather in ways that maintain safety and minimize delay. An advanced graphical interface displays traffic patterns in two and three dimensions, under various current and projected conditions for specific airspace regions or over the entire continental United States (see Figure 4.5).

Other examples of adaptation in the transportation sector

Additional examples of adaptation initiatives undertaken in relation to transport infrastructure and services used by tourists are presented on the opposite page.

Adaptation initiative	*Example*	*Source*
Climate proofing investment in road infrastructure projects	A step-by-step methodological approach for adaptation to climate change in transport sector investment projects, with a focus on road infrastructure.	ADB (2010)
Assessing the impact of climate change on transport infrastructure protected by coastal and estuarine defences	A methodology is presented for assessing the ability of coastal and estuarine defences to reduce the impact of climate change on the safety and performance of railway assets; the methodology is also applicable to the assessment of other infrastructure assets vulnerable to climate change.	Rail Safety and Standards Board (2008)
Synchronizing bridge with large vessels, tides and water levels, United States	Air gap sensors have been installed on the Verrazano Narrows Bridge at the entrance of New York Harbour, and at other locations, to ensure safety of bridge clearance by providing real-time data to port operators; the need for such information will increase due to climate change, especially as the size of vessels increases.	IFC (2009)
Increasing the resilience of sea ports to climate change, United States	Describes the steps taken by the Port Authority of New York and New Jersey, the Ports of Seattle and Corpus Christi and the Georgia Ports Authority in planning for climate change adaptation.	USEPA (2008)
Airport Adaptation	ICAO Briefing paper on climate change planning for airports.	Gittens (2010)
A reliability assessment of railway track performance in extreme heatwave events	Presents a reliability assessment of railway track buckling under extreme heatwave events – such assessments are needed for future planning and management of railway network services.	www.nccarf.edu.au/conference2010/wp-content/uploads/Minh-Nguyen.pdfwww.nccarf.edu.au/conference2010/archives/1071
Assessing the vulnerability of infrastructure	Describes an increasingly common approach to assessing the vulnerability of a structure by predicting the decrease over time in the return period of the current 1 in 100 year event that is relevant to the design of the structure.	Parkman (2008)
Sector resilience planning	Describes the UK's Sector Resilience Plans, focusing on the present level of protection of national infrastructure from flooding.	Parliamentary Office of Science and Technology (2010)

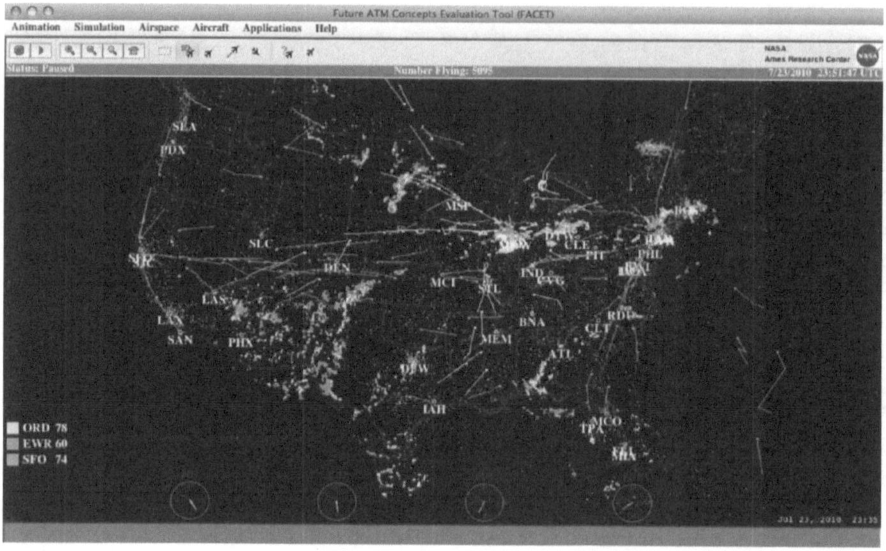

FIGURE 4.5 A real time FACET graphical display showing convective weather around Chicago O'Hare International Airport on July 23, 2010. FACET integrates weather and air traffic information, allowing operators to reroute flights to maintain safety and minimize delay. (NASA, 2010)

4.4 Managing tourism's resource base

Climate change is now another challenge to sustaining the resource base on which tourism depends. This includes not only natural resources such as water, land, ecosystems and the atmosphere, but also human resources such as culture, and production systems like vineyards. For example, Simpson *et al.* (2011) describe how climate change will directly affect the natural resources on which much of the Caribbean tourism industry is based, including the region's biodiversity and beaches. While coral reefs are a particularly important tourist attraction in the Caribbean, they are also of indirect significance to the industry. They play an essential role as physical barriers to ocean waves and as a vital habitat for fisheries.

Observed and anticipated consequences of climate change

In many areas, including the Mediterranean, tourist water demand is already starting to result in problems of overexploitation and depletion. The land area surrounding the Mediterranean Sea has experienced ten of the 12 driest winters since 1902 in just the last 20 years (Figure 4.6). The change in wintertime Mediterranean precipitation towards drier conditions between 1902 and 2010 cannot be attributed to natural variability alone. Rather, human-induced changes are considered to be responsible (Hoerling *et al.*, 2011).

Numerous studies also indicate a high probability that, for many major tourism destinations, water availability will be reduced as a consequence of both increased temperatures and decreased and more variable precipitation. Managing the resource now and in the future is constrained by the limited information on water demand and management associated with tourist consumption, especially for destinations with limitations on water supplies, such as occurs in many arid and coastal locations (Lehmann, 2009). Tortella and Tirado (2011) summarize the few studies that are available. These suggest that per capita tourist water consumption is between two and three times the local water demand in developed countries and up to 15 times in developing countries. Seasonal changes in water demand can exacerbate the supply problem, especially when they coincide with seasonal peaks in consumption by other sectors, such as agriculture. Water consumption rates also vary with the type of lodging and with the tourist activities that are undertaken, such as yachting, golf and swimming.

Coastal tourism is strongly dependent upon natural and cultural resources. While there are no reliable data on the importance of coastal tourism alone, it is generally considered to be one of the fastest growing forms of tourism in recent decades (Plates 4.3 and 4.4). In 2000, 12 of the 15 world's top destination countries in 2000 had coastlines (UNEP, 2009). Houston (2008) provides detailed information on the importance of coastal tourism to the United States, which is second only to France in international tourist arrivals, and first in terms of income

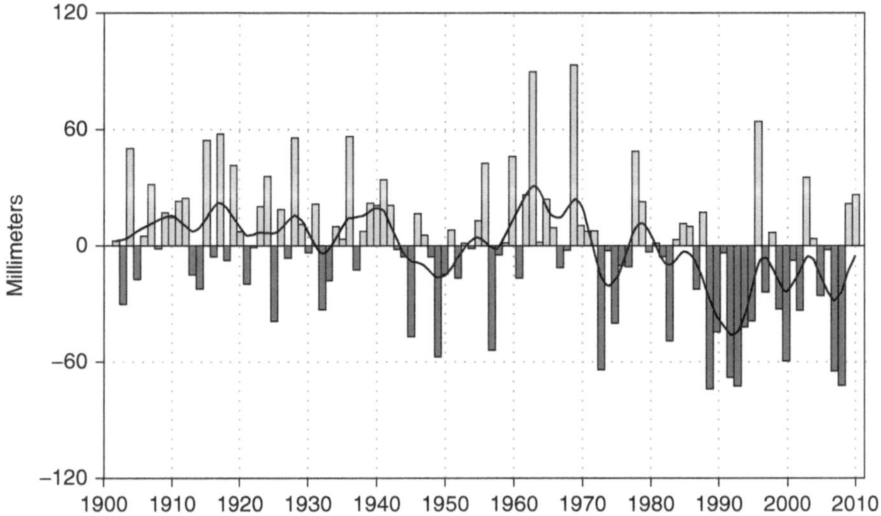

FIGURE 4.6 Winter precipitation trends in the Mediterranean region for the period 1902–2010, shown as departures from the long-term average. (Hoerling *et al.*, 2011) © American Meteorological Society. Reprinted with permission

PLATE 4.3 and

PLATE 4.4 Coastal tourism is important to the economy and livelihoods, but also vulnerable to climate change, whether in part of a large country, such as Santa Monica, United States (top) or a small island, such as Aitutaki, Cook Islands (bottom). (J. E. Hay) from international tourists (UNWTO, 2011). Its coastal states receive about 85 per cent of the tourist-related revenues in the United States, with some 180 million Americans making two billion visits to ocean, gulf and inland beaches annually. This is more than twice the number that go to all National Park Service properties combined.

Along with water, scenic beauty, rich terrestrial and marine biodiversity and diversified cultural and historic heritage, beaches form part of the unique resource mix on which coastal tourism is based. United States beaches contribute more than US$320 billion annually to the national economy. But in the United States there are over 37,000 km of eroding shoreline, with 4,300 km at a critical level. Not surprisingly therefore, erosion is the principal concern that beach tourists in the United States have about their beaches. Declining beach conditions impact adversely on tourism revenue. But conversely, in areas where eroded beaches have been restored, tourist visits and revenues have increased. For example, following the restoration of Miami Beach visits increased by 162 per cent and the annual contribution of tourism to the local economy rose to US$11 billion, including almost US$5 billion from international tourists. For every $1 the US Government invests annually, it receives $320 in tax revenues from beach tourists (Houston, 2008).

As naturally dynamic landforms, the physical condition of a beach is highly sensitive to changes in sea level, storm and wave climates, and sediment budgets. Through these processes, climate change will impact especially strongly on soft shorelines such as beaches and sand dunes, leading to accelerated coastal retreat and beach loss in these areas. Beach erosion and recession are already affecting 70 per cent of beaches globally (Bird, 2000). Zacharioudaki and Reeve (2010) investigated how changes in wave climate between a baseline (1961–1990) and a future (2071–2100) time-slice would influence shoreline evolution. They found statistically significant changes in near-shore wave climate conditions and beach alignment, with wave direction rather than wave height having the greatest influence. This finding is significant to coastal management, especially with respect to shoreline restoration and protection technologies (Plates 4.5 and 4.6). Relevant technologies are assessed in Linham and Nicholls (2010).

Climate change will also have dramatic and significant consequences for socio-ecological systems. Substantial negative impacts will occur as the global mean temperature reaches and exceeds 2°C above the pre-industrial mean, especially in biodiversity hotspots (Warren *et al.*, 2011). Terrestrial biodiversity may be especially affected, including through species shifting ranges, changes in the timing of biological cycles, modification of the frequency and intensity of wildfires, and migration of pests, diseases and invasive species. For example, based on supporting evidence from experiments, observations and modelling, climate change is identified as possibly the most significant threat facing South Africa's Cape Floral Region World Heritage site over the next 50 to 100 years. Importantly, climate change is already impacting on the Region's biodiversity (Bomhard and Midgley, 2005).

PLATE 4.5 Traditional tourist accommodation in Samoa is being threatened by accelerated coastal erosion. (J. E. Hay)

PLATE 4.6 Artificial (i.e. man-made) hotel beach at Coconut Bay, Tobago; the beach is approximately 200 m in length and 60 m in width, and was constructed using limestone sand imported from Guyana. Note also the groyne, constructed in an attempt to slow down the rate of sand loss. The developer of the hotel, John Jefferis, thought of the concept and subsequently designed and constructed the beach. (R. Mahon)

The Region consists of eight protected areas covering 553,000 ha and is characterized by an outstanding plant diversity, density and endemism.

Glick *et al.* (2011) provide a comprehensive review of the recent major trends in climate change adaptation planning and implementation, in the context of ecosystem and biodiversity conservation and natural resource management. They identified a growing emphasis on identifying and implementing more specific, actionable management options. Studies related to natural resource management generally focus on the active management of specific natural resources such as forests, fisheries or water supply, often from a resource extraction or commodities perspective. On the other hand, the primary focus of studies related to biodiversity and ecosystem conservation included protected areas management (e.g. the Giant Panda's habitat, Plate 4.7). While many recent studies still focus on providing general adaptation guidance, others are now questioning and testing the effectiveness of general strategies. This includes providing specific and actionable recommendations. The review also highlighted the lower number of studies from Africa, Asia and South America, despite these being the continents where the world's most biodiversity-rich regions are located.

Climate change is also already affecting many heritage sites, both natural and cultural, due to extreme weather and climate events. In 2005 the UNESCO World Heritage Centre distributed a questionnaire survey to all State Parties to the World Heritage Convention in order to assess the extent and nature of the impacts of climate change on World Heritage properties and actions being undertaken to manage the impacts. Of the 110 responses received from 83 States Parties, 72 per cent acknowledged that climate change had an impact on their natural and cultural heritage (World Heritage Convention Secretariat, 2008). Climate change impacts

PLATE 4.7 The Giant Panda in China is a symbol of global nature conservation (e.g. the Worldwide Fund for Nature), and climate change is seen as one factor that puts the species at risk. (S. Becken)

severely on cultural heritage, resulting in irreversible damage and losses because of its age and fragility. Ancient buildings were designed for climate conditions that may no longer prevail. Extreme weather and climate events are likely to have even greater impacts in the future. The progressive loss of these heritage resources as a result of floods, mudslides and other hazards has become a major concern, partly because of the significant role that heritage plays in contributing to social cohesion and sustainable development (World Heritage Convention Secretariat, 2008).

Many marine heritage sites are tropical coral reefs whose exposure to bleaching events is increasing, possibly leading to their mass extinction. The increase of atmospheric temperature is also leading to the melting of glaciers worldwide, in both mountainous and polar regions. The migration of pests can also have an adverse impact on the conservation of built heritage. Rising sea levels and more destructive storm surges threaten many coastal sites. Conservation of archaeological resources may be made more difficult as a result of increasing soil temperatures. Climate change will also have consequences for social and cultural aspects, with communities changing the way they live, work, worship and socialize in buildings, sites and landscapes, possibly even abandoning their built heritage as a result of being forced to migrate by increasingly adverse living conditions.

The World Heritage Convention Secretariat (2008) identifies three broad categories of action that need to be taken to safeguard heritage, namely: (i) preventive actions: monitoring, reporting and mitigation of climate change effects through environmentally sound choices and decisions at a range of levels – individual, community, institutional and corporate; (ii) corrective actions: adaptation to the reality of climate change through the implementation of global and regional strategies and local management plans; and (iii) sharing knowledge, including best practices, research, communication, public and political support, education and training, capacity building and networking.

Case study: Rwenzori Mountains National Park

The Rwenzori Mountains National Park, Uganda, contains the highest mountain range in Africa, and the third highest peak (5,109 m). The mountains are the cultural centre of the Rwenzururu Kingdom. The Park, which covers about 1,000 km^2, was designated a World Heritage site in 1994. It is best known for its scenic splendour, the trekking opportunities and the diversity in habitats contained within the park (Plates 4.8 and 4.9).

The average temperature on the mountain has risen 0.5°C every decade since 1960. One hundred years ago the glaciers on the mountains covered 7.5 km^2, now they cover 1.5 km^2, a decrease of over 20 per cent. It is anticipated that the glaciers will disappear within 15–20 years. These and other changes are significant for tourism (Jones, 2011).

The receding glacier has caused crevasses to appear in the ice, blocking the trekking paths on the mountain and requiring the establishment of new trails and bridges. New ladders and other infrastructure are needed as trekkers are now

PLATE 4.8 Rwenzori Mountains National Park, Uganda. Snow cover on the mountain is already reduced as a result of global warming. (B. Ssali, Active African Vacations, www.vacation-safaris.com)

PLATE 4.9 Trail map – the trail network and associated infrastructure are being modified as a result of changes in the climate. (B. Ssali, Active African Vacations, www.vacation-safaris.com)

walking and climbing on bare rock in areas where the glaciers used to be. It is possible that, due to further glacier melting and the appearance of more crevasses, the glacial routes will have to be abandoned altogether due to the potential danger to trekkers. Alternative routes could make the climb to the peak more challenging and technical, leading to a sharp decrease in the number of visitors to the park. At the present time a relatively unskilled trekker can climb the mountain; in the future the climb could require much more technical climbing skills, which will significantly reduce the size of the market for trekking and climbing on the mountain. Should this happen there will be a significant reduction to the considerable tourism revenue presently received by the community that borders the Park. The viability of the Park itself as a tourism destination might well be called into question.

Case study: Climate change and heritage tourism in Ireland

One of the main reasons tourists choose to travel to Ireland is the richness of the cultural and natural heritage. Changes in climate will affect this heritage, with significant implications for Ireland's tourism industry. A study was undertaken to investigate what climate change will mean for Irish tourism and the heritage, coastline and inland waterways, in order to help plan for the likely eventualities of climate change and to provide recommendations to government on prioritizing action (Chomhairle Oidhreachta/The Heritage Council, 2009). The findings highlight the potential for incompatibilities between the needs of heritage and tourism and the increasing pressure to implement adaptation initiatives. For example, the construction of coastal defences and inland flood relief works could have negative impacts on aspects of Ireland's natural and cultural heritage such as the beaches, sand dunes and historic town centres. However, mutual benefits could also result, such as where the coastline is allowed to erode or is opened to coastal flooding and inundation by the sea. Such adaptation can be beneficial to natural heritage by creating new wetlands or allowing space for coastal habitats to advance inland. On the other hand, if not planned and implemented carefully, these adaptation initiatives may also result in the destruction of archaeological sites along the coast.

Therefore, it is important that hard engineering works to protect coastal tourism assets or infrastructure take into account natural coastal processes as well as any possible long-term impacts on the heritage value, and indeed on the tourism value of the area. It is also vital that such protection works do not exacerbate the impacts of climate change as this may lead to the eventual undermining and possible complete loss of the tourism asset. These situations need to be assessed on a case by case basis, including gathering detailed information on the likely consequences of not only the anticipated changes in climate, but also of the planned interventions to reduce any impacts.

Case study: Mountain area tourism in Pakistan

In October 2009 the Chitral Association for Mountain Area Tourism (CAMAT) celebrated the International Day of Climate Action in the Kalash Valley region of Pakistan. This brought together local environmental groups, students, community leaders and women's groups. The theme of the day was 'The Impacts of Global Warming on the Kalash Communities'. Many of the participants voiced concerns about the threats caused by climate change to the natural environment of the Kalash Valley communities. These communities are already experiencing longer summers, shorter autumns and winters with less snowfall. The snow on the mountains is starting to melt earlier and earlier each year. They have also been experiencing floods with greater frequency than ever before. These destroy cultivable lands, orchards, irrigation channels and the Kalash cemeteries.

One speaker demonstrated the critical links between the health of the natural environment and the well-being of local communities by pointing out the illustrative example of the decline in the local goat population. Goats are an important part of traditional Kalash family events and religious ceremonies. The decline in goat population over the years is largely attributed to the declining forest belts, degrading alpine meadows and disappearance of vegetation cover. All these changes were attributed to global warming.

Recognizing that the core values of the Kalash culture are closely tied to healthy biodiversity and ecosystems, participants discussed the need for a comprehensive agenda focused on effective climate actions. They produced a set of recommendations including the following: initiatives that address the protection of local natural forest belts must be strengthened, and reforestation efforts must be reinforced; tourists visiting the Kalash region need to be properly informed about the fragility of the valleys, and the impacts of their activities on the springs, streams and nullahs (drains); and strict regulation of the use of non-biodegradable plastic bags is required as these negatively affect the productivity of cultivable lands – with tour companies held accountable for compliance with the regulation.

Other examples of managing tourism's resource base in the face of climate change

The following are additional practical examples of adaptation interventions targeting a particular component of the resource base for tourism.

Initiative	*Detail*	*Source*
Disaster risks and world heritage	In recognition that cultural and natural properties are increasingly affected by disasters, the manual covers the principles, methodology and process for managing disaster risks at cultural and natural sites.	UNESCO (2010)

Initiative	*Detail*	*Source*
Climate change and golf	Adaptation responses for golf courses include consulting the national governing body to seek guidance on climate change and reflecting these in course designs and maintenance plans, noting potential changes in rainfall patterns and the implications for water supply, the implications of sea-level rise (for coastal courses) and considering the impact of changes in climate on turf-grass types, water features, irrigation and pests.	www.bestcourseforgolf.org/content/administrators/planning_and_de/design_and_cons/climate_change
Saving and Making Money in a Changing Climate – A Practical Guide for Small Businesses, UK	The guide was produced to assist small and medium businesses in the West Midlands. It helps them understand how a business could be affected by weather, including extreme events, and how it can improve its chances of a quick recovery. The guide also sets out business opportunities from responding to a changing climate, and provides useful tools and contact information.	www.bitc.org.uk/document.rm?id=11938
Joint Programming Initiative on cultural heritage, climate change and security	The Initiative will mobilize the best research teams and infrastructures and their international partners active in the area of protection of cultural heritage not only in Europe but worldwide.	JPI on Cultural Heritage. Final document (4 November 2009)
Business resilience and climate change adaptation	A business resilience health check to help a business make changes to become more resilient to the risks it faces both now and in the future, by reducing its vulnerability to the effects of severe weather and ensuring it can remain in business following a severe weather event.	www.businessresiliencehealthcheck.co.uk/
Technologies for Climate Change Adaptation – Coastal Erosion and Flooding	A guide covering 13 major adaptation technologies that reduce impacts of coastal erosion and flooding due to climate change.	Linham and Nicholls (2010)

Initiative	*Detail*	*Source*
Coral Reef Alliance	CORAL has grown from a small, grassroots alliance into an international non-profit organization that works to unite communities to protect coral reefs. It provides tools, education, and inspiration to residents of coral reef destinations to support local projects that benefit both reefs and people. Climate change is an important aspect of CORAL's work.	www.coral.org/resources/issue_briefs/climate_change

4.5 Managing tourist satisfaction, health and safety

This section deals with the many aspects of tourist satisfaction, health and safety that interact with climate change. Becken *et al.* (2010a) emphasize the importance of both weather and climate for tourism, noting that the perceived conditions are often more important than the reality. Tourists make decisions based on what they believe circumstances at the destination will be, not only with respect to temperature, precipitation and other weather conditions, but also in relation to safety, for example in response to the perceived risk of tropical cyclones or other extreme events. Marketing has key roles and responsibilities to ensure tourists make well-informed decisions (Tourism Research Australia, 2007).

Responsible marketing of tourism occurs when a tourism service provider ensures that information supplied to prospective clients is a correct representation of those services. This contrasts with the marketing of responsible tourism, which is defined by SNV Netherlands Development Organisation (2009) as tourism that maximizes the benefits to local communities, minimizes negative social or environmental impacts and helps local people conserve fragile cultures, habitats or species. Responsible marketing means that, in addition to highlighting the benefits of the tourism experience, there will be full disclosure of any risks to which the client might be exposed. In the context of adaptation to climate change, responsible marketing would therefore involve making available to the client information on current weather- and climate-related risks, and on any efforts being made to manage those risks into the future.

Figure 4.7 shows that the tourism operator can be proactive in this way, but may opt for substantially less ethical marketing which, in an extreme example, makes no reference to weather- and climate-related risks. When there is a more proactive and ethical approach to marketing the tourism services, a tourist can make more informed decisions, including choice of destination, timing of travel, and choice and timing of the activities undertaken at the destination. In a changing climate ensuring that expectations match, or closely approximate, experience is a major challenge

		Communication of Weather- and Climate-related Information	
		Very Limited and Superficial	**Comprehensive and Detailed**
Weather and Climate Risk Management	**Low**	**Inactive** No support from, or involvement of, senior management Weather- and climate-related risks are assumed to be negligible No effort is made to communicate risk information No effort is made to increase employee understanding of climate change, including adaptation	**Exploitive** Some involvement of senior management Weather- and climate-related risks addressed only when beneficial Only favourable weather and climate information is communicated to potential clients Little effort to raise understanding of employees
	High	**Reactive** Senior management is involved, but only to a limited extent Weather- and climate-related risks are assessed, but not communicated to potential clients Only limited effort is made to share relevant information internally, and to build adaptive capacity	**Proactive** Assessing and managing weather and climate- risks is a priority Management involved in assessing and disseminating risk information Full disclosure of risk information, both internally and externally Employees empowered to manage weather- and climate-related risks

FIGURE 4.7 Matrix of approaches to responsible marketing of tourism services in the context of climate risks and their management. (After Hudson and Miller, 2005)

for the tourism industry. Becken (2010) has reviewed the international literature on the interactions between climate and weather on the one hand, and tourism on the other. This literature provides an insight into specific weather–recreation interactions such as warmer summers in Canada increasing the length of the golfing season.

Access to early warnings of hazards posing a threat to tourism operations is an important aspect of the ability to adapt to changes in weather and climate conditions, be they occurring on an hourly, daily or longer-term basis. Relevant early warnings include those related to heatwaves, flooding, drought, fire and disease outbreaks.

The health and safety of tourists depend partly on the ability of destinations, and individual operators, to adapt planning and management practices to address the current and anticipated impacts of climate change, including the prevention of, and recovery from, weather- and climate-related disasters. As highlighted by Robertson *et al.* (2006), there are two fundamental roles for tourism in risk management: the first is as a partner with government and community agencies in the development of multi-agency, coordinated disaster management plans, systems, procedures and processes which include the needs of tourism; the second is to develop plans and operational procedures appropriate to a destination and to the specific roles and responsibilities of a tourism business or related organization, to train personnel to the plans, and to conduct regular tests of plans, procedures and personnel. These tests would typically result in amending and updating the plans and operational procedures as well as the training of personnel.

Although the tourism industry is not responsible for the development or implementation of community disaster management plans and arrangements, destination and tourism operators should, when possible, participate in disaster planning and management activities through relevant local, regional and national committees. Such participation enables the needs and capacities of tourism to be drawn to the attention of disaster management agencies, while also encouraging that appropriate measures be established to protect tourists in times of crisis.

One of the major linkages between tourism and climate change relates to food safety. Climate change is anticipated to have both direct and indirect impacts on the occurrence of food safety hazards at various stages of the food chain. For example, rising average temperatures could increase hygiene risks associated with the storage and distribution of food commodities. Cases of salmonellosis increased by 5–10 per cent for each 1°C increase in weekly temperature, for ambient temperatures above 5°C (Tirado *et al.*, 2010). There are 38.4 million cases of food poisoning in the US every year, mostly from noroviruses. This is the pathogen best known for affecting cruise ship passengers, causing an average of 27 confirmed outbreaks annually over the past five years (Wikswo *et al.*, 2011).

These linkages highlight the need for adequate attention to food safety and security, including ensuring preparedness for the effective management of emergency situations arising from extreme weather events. Despite the uncertainty about the possible food safety implications of climate change, some governments are being proactive and are already undertaking studies aimed at identifying emerging food safety risks linked to climate change (FAO, 2009). Assuring that food is safe requires the active involvement of a number of stakeholders performing agreed, clearly defined, though necessarily interacting, roles. The various actors and actions are governed by rules – legislative measures consist of food sanitation and hygiene laws, guidelines and regulations put in place by parliament and government. The ultimate goal of this food control system is to ensure that food presented to consumers is safe and presented with integrity.

The development of relevant and enforceable food laws and regulations is an essential component of a modern food control system. Many countries have inadequate food legislation, limiting the effectiveness of all food control activities carried out in the country. Food law has traditionally consisted of legal definitions of unsafe food, the prescription of enforcement tools for removing unsafe food from use and the punishing of parties responsible for any violations. It has generally not provided food control agencies with a clear mandate and authority to prevent food safety problems. The result has been food safety programmes that are reactive and enforcement-oriented, rather than preventive and holistic in their approach to reducing the risk of foodborne illness. To the extent possible, modern food laws not only contain the necessary legal powers and prescriptions to ensure food safety, but also allow the competent food authority or authorities to build preventive approaches into the system (FAO/WHO, undated).

Case study: Tourist resort responsible marketing

The Sheraton Fiji Resort is located on Denarau Island, a major tourism destination in the South Pacific island-nation of Fiji. The resort has suffered massive erosion and consequential loss of thousands of cubic metres of the original white sand beach from in front of the resort. The erosion and loss of the main beach was due to rising sea levels and increased wave action, exacerbated by the construction of a sea wall at a neighbouring resort. This degradation of the beach has a detrimental impact on the overall satisfaction levels of guests, on opportunities for marketing and promoting the resort and on the ability of the resort to remain attractive and hence competitive (Mayes *et al.*, 2010).

Previously, as a result of cyclone damage, the resort had set back structures (Becken, 2004). As a specific response to the subsequent loss of the beach in front of the resort, three innovative response strategies were identified and implemented (Mayes *et al.*, 2010). First, to replace the original single pool and eroded dark sand beach a major renovation project was undertaken. This included installation of a much larger 1,200 m^2 lagoon-style swimming pool that extends along the beachfront grounds of the resort (Plate 4.10).

Second, guests are offered the opportunity to be ferried from the eroded home beach to a small nearby island owned by the resort where there is a more secluded and pristine white sandy beach. Third, the resort organised for its guests to have full access to the facilities of all three adjacent Sheraton and Westin resorts,

PLATE 4.10 Hotel pool on Denarau Island, Fiji. (J. E. Hay)

including use of the six swimming pools and unlimited access to the beach where the white sand eroded from in front of the resort had accumulated.

The Sheraton marketing strategy now includes proactive promotion of the innovative cluster of resorts, an emphasis on the added value resulting from three times the number and variety of activities and facilities, highlights and photos of the multi-pool complex and the offer of the unique opportunity to enjoy a half or full day 'escape' on a sandy beach on the uninhabited island (Mayes *et al.*, 2010). Significantly, while the resort's website[1] highlights the availability of full-service water sports, there are no references to the beach area in front of the resort. Such responsible marketing, which ensures a good match between expectations and experience, will help ensure high levels of guest satisfaction and repeat business.

Case study: Tourist destination with high weather and climate variability

When weather conditions deviate significantly from those considered to be ideal for the tourist, the tourism operator or the tourist can respond in a variety of ways. Based on de Freitas (2003) these are: avoid exposure to the unfavourable weather conditions, for example by changing the immediate location; offering and undertaking activities more suited to the prevailing weather conditions, such as walking on the beach rather than swimming; using structural or mechanical protection, such as an umbrella or wind break; protecting the body, for example by using additional cover; and encouraging and acknowledging acceptance of the prevailing weather conditions.

Wanaka and Queenstown are located in the South Island of New Zealand. These major tourist destinations are some 120 km apart by road. A study of the impact of weather and climate variability on their tourism businesses (Becken *et al.*, 2010a) highlights that every extreme weather event generates 'winners' and 'losers'. Such events increase operating costs (e.g. evacuation, higher maintenance costs) and are responsible for loss of revenue (e.g. cancellations). On the other hand, some operations experience increased demand when others are unable to operate. Due to the high natural variability of climate in the region, tourism businesses in Wanaka and Queenstown are well prepared for changing weather and adverse conditions. They are also geared up to make the most out of favourable conditions (Plate 4.11).

Table 4.3 summarizes the adaptation measures being used by tourism operators, as well as some of the barriers to their use. Product diversification and networking are seen as key measures to reduce the impact of adverse weather conditions. Both destinations have strong networks that allow for overall product diversification. Thus operators highlight the importance of the social fabric at their destination, with businesses helping each other as 'everyone is in the same boat'. Experience with the local climate and weather conditions was highlighted as an important business skill, as well as having skilled staff with knowledge and experience related to local weather and associated conditions. Some companies assess weather information on an hourly basis as part of normal operations. Weather stations are operated by those businesses requiring information specific to their location and operations.

PLATE 4.11 Wanaka on a sunny winter day. (S. Becken)

TABLE 4.3 Adaptation measures being used by tourism operators in Wanaka and Queenstown. (After Becken *et al.*, 2010a)

Adaptation measure	*Barriers*
Networks and diversification	
A single company increases its range of products, allowing it to move customers from one activity to another, or to borrow equipment (e.g. transport) from different business centres	High staff turnover and loss of corporate memory
Ensuring the company is highly integrated in the destination, including making a conscious effort to link with other operators – e.g. promoting 'combo passes'	Companies wanting to retain their competitive advantages
Strengthen strategic links with tour companies and inbound tour operators	Short time horizons for many business decisions
Relocating or changing the service	
Lack of snow forces heliski operators to use more limited and remote areas; fixed wing planes used to shuttle clients to the more remote staging points for heliskiing	Keeps the operation going, but costs are increased during the adverse conditions
Operators such as fishing guides, jet boat operators, hiking companies and scenic boat operators relocate their operations to more favourable locations	Relocations constrained by regulatory and other approval processes

TABLE 4.3 (*Cont'd*)

Adaptation measure	*Barriers*
Smaller rafts used during low river flows	Need additional staff as well as equipment; costs increase
Special event organisers adapt their programme to the weather conditions	Obligations to patrons to ensure the event is consistent with the advertised product
Cancellation	
A last resort action, used only as a result of extreme conditions, this may be the only option Mechanisms in place to refund tickets or compensate clients, such as by offering alternative products Booking systems adapted to accommodate cancellations and postponements during the course of a day, including rebooking to another time	Cancellations result in financial losses for the company Staff earnings adversely affected, especially for those contracted on a demand basis
Protection	
Unfavourable weather conditions may require extra protection for the tourist or the equipment, including installation of radar technology on aircraft to ensure greater reliability, adding sand bags to gondola payload during times of strong winds, the provision of weather-resistant clothing for clients, and installing rain protection on boats Some companies use different equipment as weather conditions change, such as 4WD vehicles when roads are affected by snow, ice and flooding	Need for commitment, and information that will allow well informed decision making, including extreme weather forecasts targetted to end user requirements

However, a major barrier to businesses adapting to extreme weather events, and related conditions, is a result of the low frequency of such conditions. This makes it difficult to prepare for such events, especially when costly infrastructure is required and funding is an issue. In addition, the fact that tourism operates alongside other industries may also constrain the ability to adapt. An example is competition over equipment and staff. This tends to reduce adaptive capacity.

Case study: Use of early warnings to reduce impacts on tourism

High winds and the potential for severe flooding and other hazards make it unsafe and uncomfortable for tourists to be in an area being impacted by a hurricane. Fortunately, the modern-day forecasting of tropical cyclones (also called typhoons or hurricanes) usually provides ample opportunity to safeguard people and property. Nevertheless, as Figure 4.8 shows, the impact of such an extreme event on the tourism industry can be substantial.

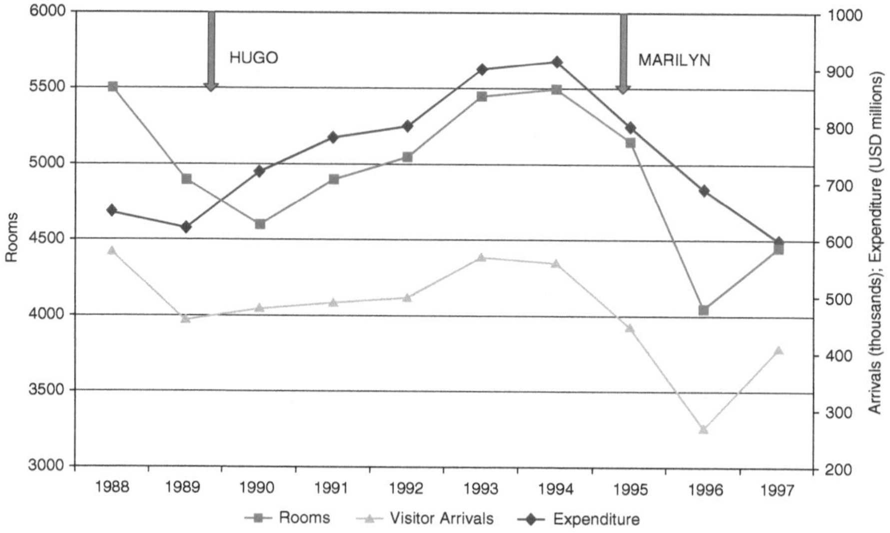

FIGURE 4.8 Impact of Hurricanes Hugo and Marilyn on room capacity, visitor arrivals and tourist expenditure in the United States Virgin Islands, 1988 to 1997. (After Caribbean Tourism Organisation)

For an average week during hurricane season, the probability of a visitor to the Florida Keys, United States, needing to evacuate because of a storm is less than 2 per cent. Nevertheless, the Monroe County (Florida Keys) Tourist Development Council has a formal communications programme to provide crucial information to help visitors safely exit the Florida Keys in the event that a hurricane threatens the region. The Council's visitor safety programme has served as a model for other destinations in hurricane-prone regions. In 2007 both the US National and Florida Hurricane Conferences honoured the Keys programme with public awareness awards.

Throughout the hurricane season emergency management officials in the Keys are in constant touch with the National Hurricane Center, based in Miami. In the event that a hurricane threatens the Keys, the communications programme is activated, in cooperation with emergency officials. Local media reports and official messages to accommodation facilities provide updates and details about possible evacuation orders. The Council also provides relevant storm-related information via a telephone hotline, its website and its Facebook and Twitter accounts.

Visitors are asked to leave the Keys when a storm of any category threatens, while residents are mandated to leave during a hurricane of Category 3 or higher. Because of the Keys' unique geography, as a chain of islands with one highway in and out, the region requires more time to evacuate than other coastal areas.

In addition, emergency officials must distribute warnings early, in order to avoid the potentially adverse repercussions of evacuations from other South Florida communities. Although emergency officials err on the side of caution, evacuation orders are only given if there is a significant storm threat.

If a visitor evacuation order is necessary, officials strive to issue it early in the morning to provide ample time for visitors to make alternative travel arrangements. The Council surveys transportation companies to determine what transport is available, and passes this information on, via advisories to accommodation facilities. The information is also made available on the website of the Tourism Development Council. The Council also works cooperatively with other Florida destinations that typically set up special hotlines to provide hotel availability and rates in the event of an evacuation order. These numbers are published in advisories sent to properties and posted on the Keys Website (Figure 4.9).

Case study: Food safety on waterborne transport in the United States

The United States' waterborne transportation industry consists of a broad range of passenger vessels, including commuter ferries, dinner cruise ships and luxury cruise ships. These carry 900 million passengers each year. Food service on these vessels varies from pre-packaged drinks and snacks, to freshly prepared gourmet meals, to buffets open 24 hours a day, seven days a week. Under the Federal Food, Drug, and Cosmetic Act, the Public Health Service Act, and the Passenger Vessel Services Act, the US Food and Drug Administration (USFDA) is responsible for enforcing food safety regulations on these vessels. It conducts operational sanitation inspections, based on the US Food Code 2009 (USFDA, 2009) for all food spaces. The inspection also covers maintenance of potable and non-potable water systems, plumbing, ventilation, waste handling and vermin control. Several regulatory options are available if a violation occurs, including issuing warning letters, seeking a temporary restraining order or permanent injunction, and collaborating with the US Coast Guard to obtain a 'no sail' order that prevents the vessel from leaving port with passengers.

Davidson and Kummer (2008) describe the additional protective measures that are now needed to control current and emerging public health threats, including the norovirus infection, salmonella and campylobacter, as well as airborne pathogens causing Legionnaire's disease, SARS and, potentially, avian flu. Efforts to control gastroenteritis outbreaks on cruise ships need to address all possible modes of norovirus transmission, including foodborne, environmental presence and persistence, and person to person spread. Successful risk reduction measures include sound food and water handling practices, extensive disinfection, isolating ill persons, providing paid sick leave for unwell crew members, and promoting hand washing with soap and water followed by the use of a hand sanitizer.

Developing strategies and incentives to dissuade symptomatic passengers from boarding vessels may also minimize the likelihood of introducing new strains aboard. Consistently and proactively applying preventive measures results in far fewer

Keys Tourism Advisory Wednesday, August 11, 2004 10.40am	
This advisory is based on information and instructions from Monroe County Emergency Management and the National Hurricane Center	
Storm Name	***Storm Category***
Charley	Tropical Storm
Current Position (Latitude; Longitude)	16.9 N; 74 W
Current Distance from Florida Keys	***Direction/Speed***
650 miles to the Southeast	WNW 17 mph
Current Status of Storm Warning	***Current Evacuation Status for Visitors***
Hurricane Watch Hurricane conditions possible within 36 h	Mandatory Evacuation for visitors and non-residents
Current Evacuation Status for Keys Residents	
No resident evacuation orders in force	

Synopsis:

Monroe Counry Emergency Management officials are ordering a non-resident and visitor evacuation for people visiting between Carig Key (mile marker 72) and Key West (mile marker 0), effective 11am Wednesday, due to a threat from Tropical Cyclone Charley. A visitor evacuation order for the remainder of the Keys may be issued later Wednesday, officials said.

Vacationers with immediate plans to visit any area in the Florida Keys should postpone their trips until the threat passes.

The National Hurricane Center has posted a hurricane watch for the Middle and Lower Florida Keys through the Dry Tortugas. The watch area begins at Craig Key and continues through Marathon, Big Pine Key and the Lower Keys, Key West and the Dry Tortugas.

A watch means that hurricane conditions are possible in the designated areas within 36 hours.

State and county parks will be asked to close later today. All recreational and high-profile vehicles should leave the Keys.

The storm is moving quickly and is likely to reach hurricane strength later today.

Preparations to secure residential and commercial property within the watch area should be rushed to completion and marine interests should secure their vessels.

Those in the Upper Keys (Kay Largo and Islamorada) should pay close attention to future advisories in the event watches or warnings are extended to that region.

Monroe County schools are open Wednesday, but schools will close Thursday and Friday.

In order to issue an evacuation order, Monroe County officials have declared a local state of emergency.

Additional advisories will be issued later Wednesday.

<u>Infrastructure</u> An "X" indicates open and/or operating normally in most or all regions at this time	X	US Highway 1	X	Marathon Airport		County Parks	<u>Future Travel to Keys?</u> Postpone travel until threat passes
	X	Electricity	X	Key West Airport		State Parks	
	X	Water	X	Key West Port	X	Hospitals	

FIGURE 4.9 Example of tourism advisory related to a hurricane approaching the Florida Keys. (Adapted from http://www.fla-keys.com/img/TourismAdvisory.gif)

illnesses than trying to contain an outbreak once it has begun. Importantly, even small gastroenteritis outbreaks on vessels should be considered for investigation because they may yield substantial information on the transmission and epidemiologic characteristics of the pathogens involved. This information can be used to develop control strategies and prevent future outbreaks on sea and land.

Case study: Tourism crisis management planning

The Queensland Tourism Crisis Management Plan (Q-TCMP), a collaborative effort involving industry and government stakeholders, covers any potential tourism crisis that will have a direct or indirect impact upon the industry. Examples of events with the potential to directly generate a crisis include natural disasters in tourism areas (e.g. cyclones, flooding, bushfires, oil spills), a tourism related industrial dispute or business collapse (e.g. pilots' strike, airline collapse) and terrorism involving tourism infrastructure or tourists. Events that can cause a crisis indirectly include health scares (e.g. foot and mouth disease, SARS), economic downturns (e.g. the Asian financial crisis) and major currency fluctuations. When faced with such crises a coordinated and consistent approach can reduce both the short- and longer-term consequences.

The Q-TCMP has three levels of activation (Figure 4.10). Different stakeholders are involved at each level of activation – Tourism Queensland Corporate Communications at level one; Tourism Queensland and the Tourism Strategy Division are involved at level two; a whole-of-government response group is activated for level three. Regional shocks that have a direct or indirect impact on the industry activate both a regional crisis management plan and the Q-TCMP. If the

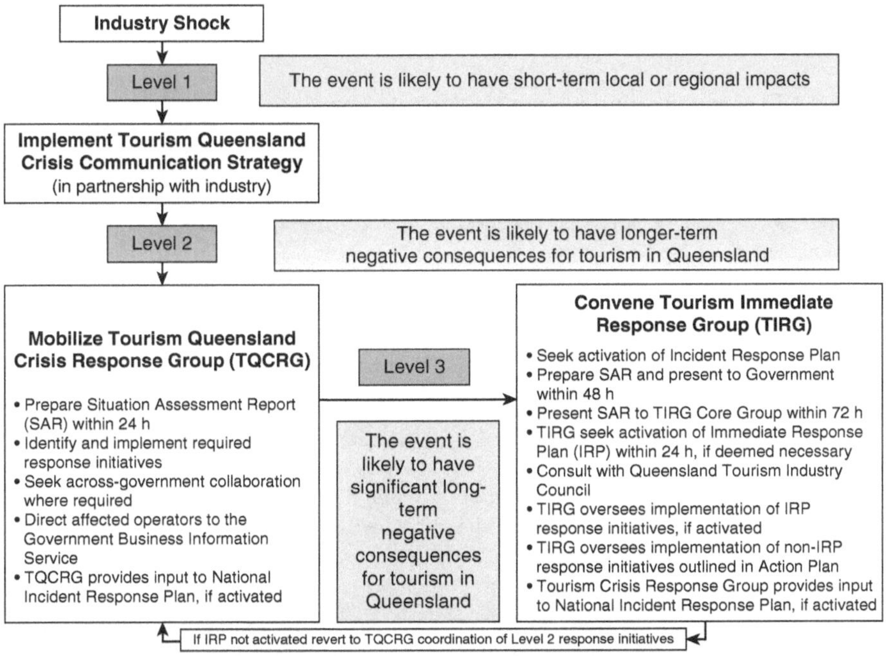

FIGURE 4.10 Queensland Tourism Crisis Management Plan. (After Tourism Queensland, 2007)

shock has short-term local or regional impacts, the Q-TCMP communication strategy will be activated. Escalation of the Q-TCMP beyond communication activities is dependent upon whether the shock is likely to reduce domestic and/or international travel to Queensland.

Since many events, such as mini tornadoes and localized rain storms, impact on regional (sub-state) tourism it is important to have a hierarchy of crisis management plans, from national through regional to local levels including the individual tourism operator level. These plans can ensure adequate preparation for, and timely response to, crises. Such plans can reduce impacts and improve the speed and extent to which businesses and regions recover from these incidents. To achieve consistency and coordination in preparedness, response and recovery activities the Q-TCMP directly links with the Australia Tourism Incident Response Plan, with regional tourism crisis management plans and with District, Regional and Local Disaster Management Groups established under the Disaster Management Act. A regional tourism crisis management plan aims to ensure a consistent and coordinated approach to response and recovery activities within the region, and between state and regional tourism bodies, in the event of a significant event impacting tourism in a region.

Other examples of managing tourist satisfaction, health and safety

Three additional examples of studies that inform and demonstrate the management of tourist satisfaction, health and safety in a changing climate are compiled below.

TABLE 4.4 Managing tourist satisfaction, health and safety.

Management focus	*Description*	*Source*
Disaster risk management for coastal tourism locations	A practical handbook to support coastal tourism destinations to prepare and respond to natural disasters.	UNEP and CAST (2008)
Responses to coral bleaching events, Australia	Highlights the need for timely and accurate communication of information on the occurrence of bleaching events.	The Great Barrier Reef Marine Park Authority Bleaching Response Plan: www.gbrmpa.gov.au/__data/assets/pdf_file/0019/4285/gbrmpa_CoralBleachingResPlan2011.pdf
Advice on planning and running a tourism business, England	Provides advice on health and safety issues.	www.tourismsoutheast.com/development/health-and-safety

4.6 Managing the sustainability of facilities, operations and destinations

Important tourist destinations are already experiencing the consequences of climate change, while many more are planning and taking action to reduce the anticipated impacts. For example, many cities in the United States that are important for tourism are increasing their readiness for climate change: San Francisco, Seattle and Phoenix are preparing for potential water supply shortages by developing local and alternative sources. These include grey water, stormwater and recycled water (Dorfman and Mehta, 2011).

Tourism destinations, and the facilities and operations within them, have lifetimes of decades and more, as highlighted in the tourism development path concept proposed by Butler (1980). Through its lifetime, a tourist area progresses from exploration through development, consolidation and stagnation, to either rejuvenation or decline. The model has been subsequently strengthened and tested by Prideaux (2000, 2004), who found that the refined model satisfactorily explained the growth path of tourism development on Australia's Gold Coast over the last 130 years.

This understanding of the life cycle of tourism has two important implications in terms of a changing climate. First, tourism assets and operations will be designed for one climate regime, but will exist and operate in increasingly different weather and climatic conditions as time passes. Second, opportunities for accommodating changes in climate will come and go, so it is important to understand when and how adaptation might best be undertaken for tourism assets and operations (Becken and Hay, 2007).

The need for a longer-term perspective on managing the sustainability of facilities, operations and destinations in light of climate change is highlighted by Simpson *et al.* (2011). A sea-level rise of 1 m is projected to put 266 out of 906 Caribbean tourism resorts and 26 out of 73 airports in the Caribbean at risk of inundation. An estimated 49 per cent of major tourism resorts in CARICOM (the Caribbean Community and Common Markets) would be damaged or destroyed by the combined effects of sea-level rise and a storm surge, in part as a result of enhanced erosion. This is because many of these resorts lack extensive coastal protection capable of preserving the aesthetics of natural beach areas and views to the sea (Plate 4.12). Simpson *et al.* warn that the longer time frames of sea-level rise should not be an excuse for delay, as putting effective protection measures in place takes considerable time – as much as 30 years or more (Hallegatte, 2009).

The following case studies illustrate how the resilience of tourism facilities, operations and destinations to climate change can be enhanced through initiatives that are sometimes described as 'climate proofing'.

Case study: 'Climate proofing' the Glastonbury Festival site

In recent years managers of the Glastonbury Festival site in South West England have noticed a significant increase in extreme weather, particularly in incidents of

PLATE 4.12 A recent trend in the Caribbean is to use white rock when constructing groynes and other forms of coastal protection, in order to improve the aesthetics while also endeavouring to ensure the beach has adequate sand cover. (R. Mahon)

heavy rainfall. This has led to drainage problems on the site, resulting in a number of severe flooding events. For example, the Glastonbury Festival in the summer of 2005 gained national media coverage for the extensive floods that affected the site, leaving thousands of tents submerged in up to six feet of water. That event was termed a '1 in 100 year storm event', yet Glastonbury experienced six such events in that year alone.

This type of event, which is expected to occur more frequently as the climate changes, has considerable impact on the business. Perceptions and experiences of bad weather can affect people wanting to come to the Festival. Flooding has made it essential for the managers to invest in improving the drainage infrastructure on the site.

The Festival organizers consider adaptation to be essential, otherwise it would not be possible to continue the event. If large-scale flooding frequently recurred, the event licence would not be granted. Since the old drainage pipes on site were inadequate to cope with the quantities of water, much bigger pipes have been installed, tripling the drainage capacity. The size of some culverts has also been increased. Hollows where water tended to collect have been filled in.

Drainage ditches are now checked on a regular basis, and kept clear of debris, particularly before a festival. Some roads running through the site have been lowered in order to prevent water being trapped. Additional roads, both temporary and permanent, have been built in order to improve vehicle access regardless of ground conditions. A hard bus-pad has also been added, to provide space for buses to turn around. Festival tents are now held down securely in case of windy weather.

The Festival maintains a Silver Command System, to keep a close watch on the weather. This system mirrors that used by the police force. Ahead of the Festival, customers are sent advice to come prepared in case of bad weather. During the Festival trained welfare officers and stewards are on-hand to assist people in the event of extreme weather – whether this is heatwaves, intense rainfall or storms.

Case study: Water use by hotels in Mallorca, Spain

Based on a combination of global climate models and downscaling studies, Calbó (2010) concludes that it is highly probable water availability will be reduced in the Mediterranean areas as a consequence of both increased temperatures and decreased and more variable precipitation. Tortella and Tirado (2011) provide detailed insights on how a combination of management policies and guest behaviour can influence water consumption rates in tourist accommodation on Mallorca (or Majorca), Spain. The island is a popular destination for tourists from Britain, Germany, Ireland and the Scandinavian countries. The importance of the island as a tourist destination has been steadily growing since the 1950s, with 8.9 million international tourists and 2.6 million domestic tourists in 2010. Mallorca receives 76.3 per cent of all tourist arrivals during the six-month period from May to October and 56.4 per cent from June to September.

The findings are based on a survey of 200 of the 748 hotels in Mallorca. Water consumption (in litres per guest night sold, or l/gns) ranged from 156 to 2,425, with a mean value of 541 l/gns. The average consumption of a three-star hotel was 516, while consumption in four- and five-star hotels is higher, at 548 and 701 l/gns, respectively. Hotels that differentiate their product from those of their rivals, in order to achieve monopolistic rents and increase their profit margins, consume 20 per cent more water than those hotels that try to increase profit margins by following a cost reduction strategy. Hotels focusing on the differentiation strategy usually introduce new services (e.g. indoor swimming pool, sauna, jacuzzi) and higher cleaning and other quality standards. On the other hand, hotels following a cost-reduction strategy try to reduce the cost of water and other consumables, while also reducing both the services they offer and the hotel quality standards.

Importantly, as mature destinations such Mallorca Island move away from the traditional sun and sand mass tourism model to higher-quality models, water consumption will increase. While changes to a higher standard of hotel usually result in a reduced number of rooms, thereby decreasing water consumption, this effect is more than offset by the new services (e.g. swimming pools and golf courses) and higher cleaning and other quality standards. In addition, high-quality hotels usually stay open for longer, albeit with lower occupancy rates, again increasing water consumption.

The type of board (or meal plans) offered by hotels also has a significant effect on water consumption, with each additional meal increasing annual hotel water consumption by 10 per cent. The more meals a hotel serves, the more time guests spend in the hotel rooms and facilities, leading to higher water consumption by guests, both directly for ablutions and drinking water, and indirectly for water use in kitchens and laundries. Thus the introduction of all-inclusive packages increases water consumption, per guest and overall.

Tortella and Tirado confirm the effectiveness of water saving initiatives undertaken by hotels, with such initiatives reducing annual average consumption by up to 13.6 per cent. The main factors influencing hotels to introduce such measures are primarily demand-based, including improving the hotel's image, improving the quality of service, and increasing customer fidelity. The greater the importance given to environmental issues in the hotel marketing policies, the higher the probability of water saving initiatives being introduced. On the other hand, existence of an environmental department in the hotel, recognition of the severity of water regulations, and water costs do not play a relevant role. Typically hotels affiliated to small chains had significantly lower water consumption (18 per cent) relative to independent hotels, while hotels affiliated to large international chains had significantly higher levels of water consumption (34 per cent).

The findings described above illustrate that, as a tourism destination matures, hotel management policies and decisions, as well as the behaviour of guests, may well be contrary to those that should be undertaken in order to reduce consumption as water supplies decrease. Thus a business-as-usual approach constitutes maladaptation. Effective adaptation would require the increased use of measures such as those described above in order to reduce water consumption. This way it may be possible to more than offset the increased consumption resulting from management policies related to product differentiation strategies.

Case study: 'Climate proofing' tourism facilities

Since Newquay Zoo opened in 1969 it has become one of the UK's top zoos, and one of the major tourist attractions in the South West of England. In 2009 serious flooding occurred in the Oriental Garden, one of the Zoo's enclosures. The rapidly rising flood waters caused concerns not only for the welfare of the animals, but also for the integrity of the enclosures. This, and other similar flood events resulting from heavy rainfall, resulted in a decision to improve the zoo's resilience to flooding, especially since heavy rainfall events are likely to become more frequent and intense due to climate change.

Adaptation interventions involved re-landscaping the otter enclosure in the Oriental Garden, including construction of a new bridge and granite stone wall to solve the problems of persistent flooding. Enclosures have also been raised above normal flood level and a Flood Plan has been developed for staff, particularly those in charge of the most commonly affected enclosures. The Plan ensures that all staff know what to do in the event of a flood, including implementing short-term measures such as laying down sandbags and moving animals to safety. In addition to the flood resilience measures, the Zoo is adapting by ensuring that it conserves water wherever possible. Water supplies will be put under increasing pressure due to droughts and reduced summer rainfall. Water usage is now monitored and water efficient taps have been installed. Other water efficiency measures include obtaining an abstraction licence so water from the Zoo's lake can be used for operational purposes such as washing down enclosures. The Zoo plans to build a well to access spring water, which will also be used for washing down enclosures (Climate Southwest, 2010).

Case study: Climate change creates opportunities for Arctic cruise ships

One approach to adaptation is to take advantage of opportunities. A potentially positive impact of climate change relates to shipping routes. Specifically, rising temperatures in the Arctic are opening some new routes to shipping, including cruise ships. The number and range of cruise ships moving further north is increasing, with previously untouched coastal areas being visited (Forbes, 2011). The unspoiled natural beauty of the far north is a major factor in the popularity of the 'green' cruising movement. As new areas become navigable there will be the inevitable rush to be the first tourists into a region. The popularity of the Arctic as a cruising destination will only increase as the media focuses on the changing environment and the disappearance of what is thought of as the quintessential Arctic landscape (Bruno, 2011).

By 2080 the ice-free season of the Northern Sea Route could increase by up to 80 days per year. Satellite images showed that in the summer of 2007 sea ice cover in the North-west Passage had shrunk to its lowest level since 1978, when satellite measurements began. Many experts expect the Arctic to be ice free before the date projected by the IPCC (i.e. mid-2070). Recent satellite images show that open water now stretches all the way round the Arctic, making it possible to circumnavigate the North Pole for the first time in human history (UNCTAD, 2009). However, Stewart *et al.* (2007) argue that suggestions of an ice-free Arctic as early as 2050–70 may lead to a false sense of optimism regarding the potential exploitation of all Canadian Arctic waters for tourism purposes. This is because the changing character and distribution of sea ice will increase the likelihood of hull-penetrating multi-year ice that could impede Arctic navigation.

Some cruise ships have made Arctic transits, and destination cruises are now commonplace (Plate 4.13). The number of cruise ships calling into Greenland has increased from 17 in 1998 to 42 in 2008 (Statistics Greenland, 2011). The demands of a sightseeing trip are somewhat less onerous than those of a scheduled cargo route. If conditions become dangerous, cruise operators in the region have the opportunity to change their route, or visit a more accessible location. Liability considerations alone keep many ships out of dangerous areas.

Given the remote and harsh Arctic environment, search and rescue operations are extremely risky. In 2010 the *Clipper Adventurer*, a double hull passenger ship,

PLATE 4.13 Russian icebreaker *Kapitan Khlebnikov* – retrofitted for passengers in Pond Inlet, Nunavut, Canada, in August 2010. (E. Stewart)

waited two days after running aground before passengers could be safely transported to land. Current emergency response assets cannot adequately provide for the safety of the increasing numbers of ships and people. Increased maritime tourism in the Arctic thus poses serious threats for crews and passengers because the cruise ships are often not ice-strengthened and, in many cases, lack charts and adequate knowledge on Arctic conditions. Search and rescue systems do not exist in most parts of the Arctic. Even where systems are in place, they are likely to be insufficient for dealing with the large number of passengers on a cruise ship (Jariabka, 2008). While safe navigation should rely on experienced and well-trained personnel, at present there are no mandatory international ice navigation training standards (Des Bois, 2011).

A requirement for ships to go in pairs could be one regulation that is effective in increasing safety. In cases of entrapment, or mechanical failure, this would allow passengers to be transferred to another vessel, rather than waiting for a response from one of the few rescue vessels available in the area. However, even the logistics of passenger transfer are problematic because entrapment or grounding of a vessel may not allow for a second ship to take up a rescue position for ship-to-ship cable transfers. Hardened tenders are one solution, but small craft will not be able to enter heavy ice jams. Lifeboat design is another consideration for Arctic travel. Enclosed lifeboats are considered best practice for extreme conditions. This is a significant expense for an operator that is refitting a vessel for Arctic cruising. Changing transport and trade patterns are also likely to affect infrastructure investments. Ports and terminals in the Arctic need to be able to berth ice-class ships, equipment needs to be sturdy and adequate and labour needs to be skilled and specialized.

An increase in cruise-ship tourism is one of many impacts of climate change being experienced by the people of Ulukhaktok, a small community of primarily Inuit people in the Inuvialuit Settlement Region of the western Canadian Arctic. Climate change is considered a significant challenge for Inuit communities. As noted above, some impacts are already being experienced. Increased cruise ship traffic can boost tourism revenue for local communities, but may also cause problems if ships are unregulated, for example with respect to the discharge of sewage and other contaminants (Pearce *et al.*, 2010). Stewart *et al.* (2010) also point out that, while warming temperatures may facilitate both earlier and later shipping in the Hudson Bay region of Canada, reduction in ice coverage may in fact lead to a decline in cruise activity in most parts of that region. This is because ice-supported wildlife will likely shift north in response to the diminishing ice cover.

4.7 Conclusion

This chapter has documented the considerable action and learning now taking place within the tourism sector as it responds to both the current and anticipated changes in climate. The parts of the tourism system where the main effects of climate change are likely to be experienced have been identified. In some instances initiatives are already being undertaken to reduce the adverse consequences, or take advantages of the benefits. Such initiatives involve a combination of planning and tangible, on-the-ground measures. Not all components of the tourism system are responding in the same way. For example, managing tourism's resource base involves, in part, a reaction to changes that have already occurred. In contrast, current actions related to managing tourist satisfaction, health and safety are more proactive in relation to climate change. The tourism sector needs to be even more proactive in planning for climate change. This highlights the need for effective policies and strong institutions, as discussed in Chapter 3.

The practical experience that has already been accumulated by the tourism sector suggests that the following constitute good practice in managing climate-related risks and opportunities for tourism, through proactive adaptation measures, regardless of which parts of the tourism system are being considered:

- Assess the current and anticipated levels of weather- and climate-related risks to, and opportunities for, tourism operations and experiences.
- Determine when and how adaptation by tourism operators and other key players should be undertaken.
- Assess the current capacity of tourism operators and other key players to plan and implement the required adaptation measures, and what additional capacity is needed.
- Strengthen the capacity of the tourism sector to adapt, including the enabling environment for adaptation (see Chapter 3) and the ability to

monitor and evaluate the success, or otherwise, of adaptation measures (see Chapter 7).

- Implement tourism adaptation measures as and when required, and undertake appropriate monitoring and regular evaluation, in order to identify and scale up successful adaptation interventions by tourism operators and other key players, and modify those that are not fulfilling expectations.

5

MITIGATION POLICY

5.1 Introduction

'Mitigating' climate change means reducing the build-up of GHGs in the atmosphere. For tourism, this relates mainly to CO_2, which is produced by the combustion of fossil fuels. Hence, any initiative that: (a) conserves energy; (b) uses energy more efficiently; or (c) replaces fossil energy with renewable energy, contributes to the reduction of the anthropogenic greenhouse effect (Becken and Hay, 2007). The offsetting of CO_2 emissions by an investment into carbon sequestration – most commonly through planting trees or re-vegetation projects – is also considered to be a (potentially) useful mitigation measure. Policies related to mitigation will be discussed in this chapter, and examples of good practice in tourism mitigation are presented in Chapter 6.

The consideration of environmental goals in tourism policy became more prominent in the 1990s and early 2000s. Largely this was in response to growing biophysical impacts (e.g. on vegetation, water and biodiversity), and increasingly scarce resources, such as water and energy. A 2002 report on the 'Environmental Management Insertion in Tourism Sector Policies in the Caribbean' is a good example. In this report (Sustainable Economic Development Unit, 2002), Barbados, the Bahamas, Saint Lucia and Tobago are used to illustrate the state of the environment, perceptions of the environment by tourism stakeholders and the need for specific tourism policies to maintain environmental conditions and assets at the high standards that are required if a tourism destination is to be successful.

The explicit consideration of climate change as part of tourism policy dates back to the first United Nations World Tourism Organisation conference on tourism and climate change in 2003 in Djerba, Tunisia. The Djerba Declaration was the first substantial policy document on climate change by the tourism sector. From then on tourism policies were enhanced by adding climate change mitigation, or specific policies were developed by tourism stakeholders at different levels to address the GHG emissions from tourism. Private sector policy initiatives, for example, cover such approaches as voluntary initiatives or emissions targets (Table 5.1). The IPCC (2007a) provides several reasons why businesses undertake mitigation actions: (i) to influence or pre-empt government action; (ii) to create financial benefits; and (iii) to differentiate a company or its products (brands) from competitors. All these reasons apply to tourism businesses and are congruent with earlier research on why tourism operators become involved in environmental management. Examples of business policies for climate change are presented in Section 5.2.

Mitigation policies and actions by local governments and destinations, states or regions often relate to transport management and infrastructure, renewable energy portfolio standards, energy efficiency programmes, emission registries and cap and trade mechanisms. The reason for policies at the sub-national level is to inform national policies, to address the concerns of local stakeholders, to create incentives for new industries and to exploit ancillary benefits, such as those resulting from improved air quality. The latter is often a local government responsibility (IPCC, 2007b). The motivation for national level policies is multi-faceted, but may include international commitments made by the government, for example the ratification of the Kyoto Protocol. Examples of policies at local, national and regional levels are presented in

TABLE 5.1 Examples of emission reduction targets relevant to tourism.

Organisation	*Emission reduction target*	*Source*
IPCC	Recommends a reduction by 50–85% between 2000 and 2050	Scott *et al.* (2010)
World Economic Forum	-17 Gt by 2020 (no baseline year provided)	Scott *et al.* (2010)
WTTC	Aspirational target: -25% to 30% by 2020 and -50% by 2035, compared to 2005	Scott *et al.* (2010)
Europe	80% reduction in CO_2 by 2050 relative to 1990 and 20% by 2020	http://ec.europa.eu/clima/documentation/roadmap/docs/com_2011_112_en.pdf
United Kingdom	The UK Climate Change Act sets a legally binding target of reducing GHG emissions by 80% by 2050 and 26% by 2020 compared with 1990	www.nationalgrid.com/corporate/About+Us/climate/
InterContinental Hotel Group	Reduce by 25% by 2017 (no start date)	Zientara and Bohdanowicz (2010)
Hilton Worldwide	Reduce by 20% between 2009 and 2014	Zientara and Bohdanowicz (2010)

this chapter, along with business policies. The chapter concludes by examining two cases of international policy making for tourism and climate change.

> Public policy is critical to encouraging investments in a low carbon economy as it supports the adoption and development of new technologies and creates a level playing field for renewable energy investments. Policy makers must realise that nothing is as effective in discouraging investors as policies which keep changing. Targets and policies, therefore, need to be clear, effective, and long term. (David Russell, Co-head of Responsible Investment with the UK pension fund USS)

5.2 Business level mitigation policies

The tourism industry is diverse and encompasses a wide range of businesses. The World Travel & Tourism Council (WTTC), for example, represents one hundred of the world's largest tourism companies. However, the majority of tourism businesses are small to medium-sized enterprises. Often, small businesses are not represented by tourism industry associations. Notwithstanding, tourism businesses have become increasingly involved in climate change mitigation, collectively and individually. The development of numerous tourism codes, guides and policies for reducing GHG emissions is evidence of this industry engagement. These voluntary initiatives have been put in place for many reasons, including cost savings (e.g. reduced energy costs), appealing to 'green' consumers, influencing or pre-empting mandatory government policy, risk management and product development. Studies have shown that larger businesses, or those affiliated with chains, are more successful in implementing climate mitigation, partly because they have the capacity and resources to develop company policies which can then be put into practice at individual locations (Mensah, 2006).

The development of mitigation policy by businesses is often approached from a risk perspective to: (i) understand the risk; and (ii) manage it (Becken and Hay, 2007). The exposure to risks associated with increasing atmospheric concentrations of GHGs goes beyond the individual tourism business and includes risks associated with the supply chain and competitors (Figure 5.1). These different risk components of a tourism business demand different policy responses.

As Chapter 6 (Mitigation Practice) shows, there are many informative examples of climate change mitigation actions implemented by tourism businesses. It has been more difficult to identify policies or strategies that specifically address climate change. This view is confirmed by Zotz (2010) who noted that European tour operators are involved in energy efficiency and carbon reduction, but rarely develop strategies or explicit reduction targets. In contrast, corporate strategies often aim

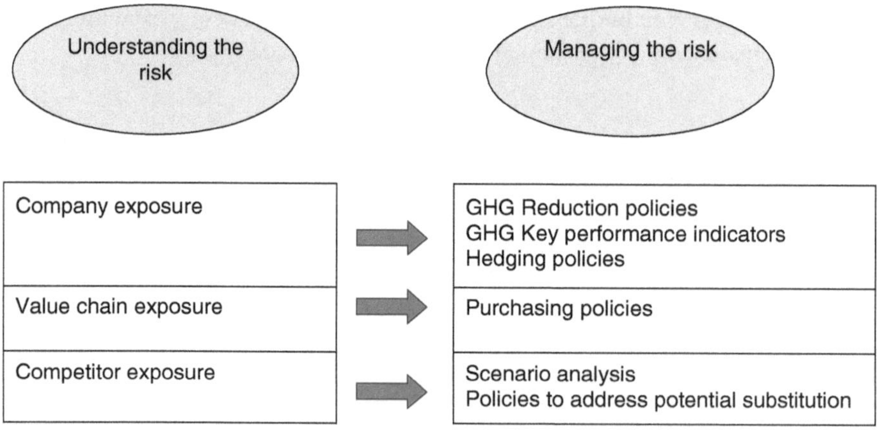

FIGURE 5.1 Risk-based framework for developing GHG mitigation policies.

at long-term growth and at enhancing long-haul segments for package tourism without addressing concomitant growth in emissions. Notwithstanding, the following case studies illustrate the variety of policies that can be found in the tourism industry. Two case studies from the accommodation sector will highlight policies designed to reduce the contributions a business makes to climate change. A third case study discusses policies implemented by a major European rail company, Eurostar.

> Reducing carbon emissions makes good business sense, just as eliminating any form of waste does. As the price of carbon and the cost of emissions rise in the future, organisations will concentrate more time and effort reducing their dependence on fossil fuel. Our economic and environmental needs will converge. (Management Consultancies Association (2007) for Arup)

Case study of a global hotel chain

Accor is a global network of hotels with headquarters in France. It is known for its leadership in environmental management, including the issue of climate change mitigation. According to the chain's 2009/2010 Sustainability Report, Accor's 2,806 facilities consumed 4,974 GWh of energy, equivalent to the domestic consumption of a city of 590,000 people. Accor emitted 1,933,000 tonnes of CO_2 as a result (Accor, 2010). Over the last three years the Accor group reduced energy use per available room by 7.8 per cent. Over 80 per cent of owned and leased hotels now use compact fluorescent light bulbs for areas that remain lit 24 hours a day. A total of 99 Accor hotels are equipped with solar-powered hot water systems.

Accor created Environment Charters to bring together its network of 4,100 hotels (owned or leased) around a common 'road map'. To allow for differences between offices and hotels, Accor has developed two specific Environment Charters. The Charters cover several areas, including energy, informing and raising awareness, water, wastewater, waste, ozone layer, biodiversity and supply chains. The energy-related items for the Hotel Environment Charter are shown in Figure 5.2. The 2010 objectives were to apply the Hotel Environment Charter in all owned and leased hotels and in the headquarters and offices of all Accor units. The 2009 year-end indicator showed that 97 per cent of owned and leased hotels had already complied with this request. A Charter poster is also displayed for guests.

ACCOR HOTELS ENVIRONMENT CHARTER – ENERGY

- Set objectives for limiting consumption
- Monitor and analyze consumption every month
- List potential technical improvements
- Organize preventive maintenance
- Ensure optimal use of plant and machinery
- Install an efficient façade lighting
- Use low energy lamps for permanent lighting
- Use low energy lamps in bedrooms
- Use LEDs (Light Emitting Diodes) for external illuminated signs
- Use LEDs for the emergency exit signs
- Use energy-efficient refrigerators in bedrooms
- Insulate pipes carrying hot/chilled fluids
- Use energy-efficient boilers
- Recover energy from the main ventilation system
- Use energy-efficient air-conditioning systems
- Recover energy from air-conditioning systems
- Install solar panels for production of domestic hot water
- Install solar panels for heating swimming pools
- Promote use of renewable energies

FIGURE 5.2 Accor Hotels Environment Charter. (www.accor.com)

In response to guidance 'from the top', many Accor hotels around the world are making substantial progress in environmental management, including GHG reduction. Recently Accor received the 2010 Tourism for Tomorrow Award from the WTTC. In addition, many individual hotels have been acknowledged for their leadership. For example, five Accor hotels in Bali, Indonesia (Sofitel Seminyak Bali, Novotel Bali Benoa, Novotel Bali Nusa Dua Hotel and Residences, Mercure Resort Sanur and All Seasons Legian) received the Tri Hita Karana Tourism Awards and Accreditation in 2009. The Sofitel Imperial in Mauritius was the winner of the Award for Sustainable Tourism in the National Quality Awards. The Sofitel

Cocagne Eindhoven in the Netherlands made history as the first hotel in the world not to have an incandescent light bulb.

Case study of a luxury resort

Six Senses was established in 1995 in the Maldives and has since become a global brand in the luxury resort sector. There are now five sub-brands – Soneva, Six Senses, Evason, Six Senses Spas and Six Senses Sanctuary. In 2009 Six Senses had 26 resorts that were either open or under development, and 41 Six Senses Spas that were in the group's own properties or hosted in properties managed by a third party. Six Senses' Sustainability Policy sets out the following objectives directly related to energy and GHG emission management that are to be met during both development and operation of their resorts:

- Address the issue of climate change through both energy management, as part of resource management policies, and avoidance of fugitive emission of chlorofluorocarbons (CFCs).
- Develop action plans as well as regularly monitor social and environmental impacts through regular environmental meetings, monitoring and updating of the Key Sustainability Indicators (KSIs) Database.
- Reduce resource consumption and waste generation through responsible waste reduction policies, reuse, recycling and composting programmes.
- Systematic management of energy use and consumption utilizing, where possible, renewable energy technologies.

The Sustainability Policy available on the Six Senses website does not provide any further detail on how the above objectives are going to be achieved, and several issues remain unclear. For example, the establishment of group-wide standards that 'wherever possible meet the local legislative environmental requirements as well as complying with international agreements' does not specify which requirements are being referred to and what the qualifying phrase 'where possible' means. Similarly, it is not clear what the specific goals of the promotion of 'sustainability awareness' are.

Despite this (not uncommon) fuzziness, the Six Senses example constitutes an often-referred to good-practice case study of sustainable management in a tourism business. The most creative innovation of Six Senses is perhaps the development of tourist experiences under the theme of SLOW LIFE 'TM' (Sustainable-Local-Organic-Wholesome Learning-Inspiring-Fun-Experiences). One specific aim is to reduce the carbon footprint by using produce that can be sourced locally. The 'Virtuous Circle' designed by Six Senses illustrates the company's interpretation of the mutually dependent relationship between sustainable practices, customer experience, corporate culture, employee enthusiasm and tourists' loyalty to the brand (Figure 5.3).

FIGURE 5.3 Six Senses Virtuous Circle. (www.sixsenses.com/Environment/Six-Senses-Sustainability-Policy.php)

Case study: Mitigation policies for train travel

The Eurostar, linking the United Kingdom with the European continent, is a prime example of the possible substitution from air travel to low carbon land transport. The train links London with Paris and Brussels, and also connects with a number of other high-speed train services in France, Germany and the Netherlands (Plate 5.1). Due to its direct city-centre to city-centre nature, it has a competitive advantage relative to airlines that require secondary transportation to and from an airport. In 2007 Eurostar launched services from its new London station, St Pancras International. It now offers travel times of two hours 15 minutes on the London–Paris route and one hour 51 minutes on the London–Brussels route.

In 2006 a commissioned study found that Eurostar journeys from London to Paris or Brussels emit a tenth of the carbon dioxide emitted had the journey been by air. The company markets itself by explicitly comparing the carbon footprints for various types of journeys. For example, it states that return travel by Eurostar from St Pancras International in London to Paris Gare du Nord emits 6.6 kg of CO_2 per passenger, compared with 102.8 kg of CO_2 per

PLATE 5.1 Eurostar in Brussels. (Eurostar Press Office)

passenger for a return flight from London Luton to Paris Charles de Gaulle. Specific policies developed to address the company's environmental and carbon footprints include:

- an emission reduction target of 35 per cent by 2012, relative to 2007;
- a 'Tread Lightly' environment programme that achieved carbon neutrality for Eurostar since 2007;
- a carbon offsetting plan;
- a ten-point plan to minimize climate impacts, for example by replacing train air-conditioning refrigerants and developing a 'switch off' culture to reduce electricity demand;
- a Sustainable Procurement Policy to ensure the reduction of environmental impacts as a result of supply chain activities; and
- driver education – Eurostar fitted one of their trains with a device that measures how much energy it is using.[1]

Other examples of tourism business mitigation policies

There are many potential case studies of tourism and travel companies that have addressed climate change in their environmental policies or that have specifically designed climate change policies. Below are a few examples to illustrate the diversity of policy initiatives.

Company	*Policy*	*Source*
Train, SNCF, France	SNCF has introduced a sustainable development aspect into each group of purchases, its purchasing specifications and in the choice of suppliers. A particular focus is on GHG emissions.	www.sncf.com/en_EN/html/page/CH0006-Ecomobility/BR1123-Award-Sustainable-purchasing.html
Asia Adventures, Cambodia	As part of 'Responsible Tourism' the company developed an Asia Adventure Responsible Tourism Travellers Code of Conduct with specific items for energy and transport.	www.asia-adventures.com/information/responsible-tourism.php
Auckland International Airport, New Zealand	Auckland International Airport has a Sustainability as well as an Environmental Policy. In 2008 a sustainability forum was established, which produces a five-year sustainability action plan that will ensure 14 key areas of sustainable business performance are monitored and managed.	www.aucklandairport.co.nz/Social-Responsibility/Sustainability-policy.aspx
TUI Travel, Europe	The Annual Sustainable Development Report details how TUI is positioning in relation to climate policies and provides information on internal carbon management programmes.	http://sd2009.tuitravelplc.com/tui-sd2009/en/carbon-management
AVIS, United States	AVIS has established 'smart company policies' for energy and other areas of concern, including SmartWay® certified vehicles.	www.avis.com/car-rental/content/display.ac?contentId=avisgreeninitiative-US_en-018201
Virgin Atlantic, Great Britain	Climate change mitigation is the number one issue in Virgin's Environmental Policy.	www.virgin-atlantic.com/tridion/images/newpolicy_tcm4-540700.pdf
British Airways, Great Britain	BA's policy details pillars of climate change mitigation: energy efficiency, emissions trading and avoided deforestation.	www.britishairways.com/travel/csr-climate-policy/public/en_gb

5.3 Local mitigation policies

Background

Local policies and initiatives are extremely important because it is at this level that many actual changes can happen, for example through the installation of more energy-efficient equipment or the construction of low carbon infrastructure (e.g. cycle ways). As a result, local policy making and planning have gained increased recognition in the last few years. Local government agencies are often at the core of decisions relating to both climate change mitigation and adaptation. In partnership with industry bodies and NGOs they represent local communities, destinations and businesses. Tourist destinations (e.g. through Regional Tourism Organizations) often, but not always, collaborate with local governments to develop policies and implement climate change mitigation measures. Tourist destinations often wish to differentiate themselves by acting as leaders of climate change policies, even if the country as a whole may lag behind. Initiatives such as the National Geographic Centre of Sustainable Destinations (and Destination Stewardship surveys), and destination-based sustainability charters, are evidence of this.[2]

ICLEI-Local Governments for Sustainability (formerly International Council for Local Environmental Initiatives) is a global body that brings together and represents local governments around the world. ICLEI's members include 1,200 local governments from 70 different countries. Many of these are important tourist destinations, for example Barcelona (Spain), Buenos Aires (Argentina), Rio de Janeiro (Brazil), Vancouver (Canada), Montpellier (France), Siena (Italy), Kyoto (Japan), and Arusha (Tanzania). ICLEI cooperates with the United Nations Environment Programme (UNEP) on the Local Government Climate Roadmap, a campaign for a comprehensive post-2012 global climate agreement with strong local participation.

The following section provides an example (the Sunshine Coast) of how climate change mitigation for tourism is explicitly included in local government strategies and plans. A second example (Kathmandu) highlights how goals other than climate change lead to climate change mitigation at a tourist destination, as well as the pitfalls such projects might experience.

Case study of local government mitigation polices

The Sunshine Coast, an important tourist destination, is just north of Brisbane in Queensland, Australia. Tourism contributes 17 per cent to the Gross Regional Product, which makes the Sunshine Coast five times more reliant on tourism compared to Queensland as a whole, and four times more than all of Australia, where tourism contributes 3.9 per cent of the Gross Domestic Product (Sunshine Coast Council, 2010a).

As part of the Sunshine Coast Council's Corporate Plan 2009–2014 (Figure 5.4), the council developed a Climate Change and Peak Oil Strategy. Key components of this strategy that are relevant to climate change mitigation are the Energy

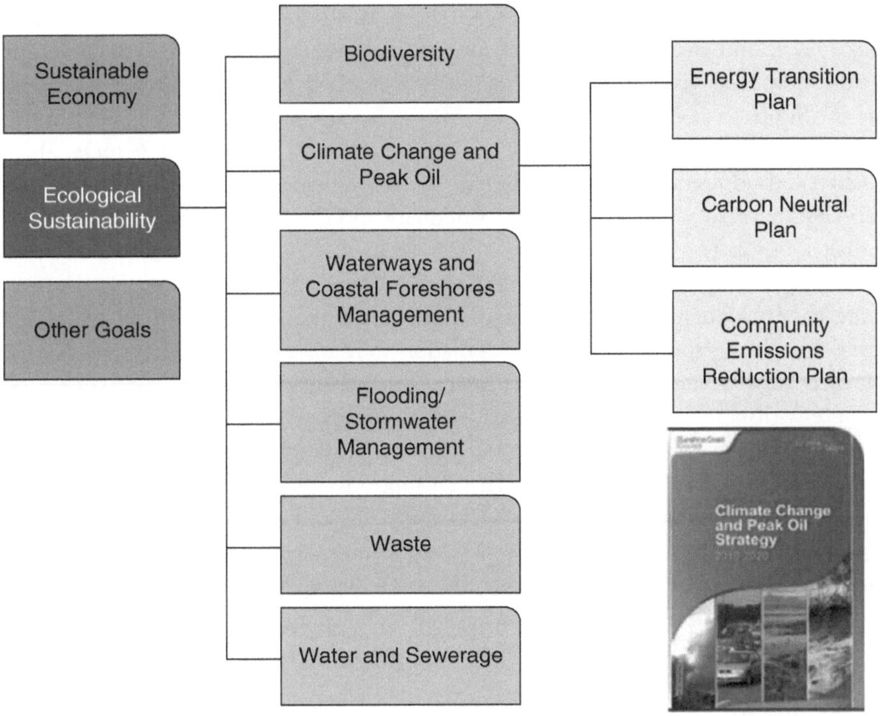

FIGURE 5.4 Corporate Plan 2009–2011 of the Sunshine Coast Council (Australia) (2010a).

Transition Plan, the Carbon Neutral Plan and the Community Emissions Reduction Plan. All three are based on the assumption of increasing (and volatile) oil prices that will be exacerbated by increased costs of carbon due to climate policies. The implications of higher energy costs were assessed in detail for five sectors: transport; construction; food supply; regional energy production; and tourism. It is noteworthy that all four non-tourism sectors provide essential inputs into tourism. Hence the analysis of all five sectors is of great importance to future-proofing tourism.

A tourism-specific risk assessment concluded that, while the Sunshine Coast's reliance on tourism is high, at this point in time its vulnerability to higher energy costs is relatively low. This is due to the very small proportion of international tourists (only 9 per cent of arrivals) who arrive on long-haul flights. However, it is recognized that many Australian tourists arrive by low-cost carriers. As a result, the assessment of airline viability is one of the recommendations made for further studies.

Case study: Linking local and global issues

Kathmandu is an important tourist destination in itself and also the main gateway for tourists to other parts of Nepal. Local air pollution is a major problem in Kathmandu, with adverse effects on both the local population and tourists. A survey found that 41 per cent of tourists had breathing problems, 58 per cent reported limited visibility, and 69 per cent described Kathmandu's air quality as 'poor' (Simpson *et al.*, 2008b: Box 30). The main reasons for poor air quality in the Kathmandu Valley were found to be 'old and poorly maintained vehicles, low quality and adulterated fuel, prevalence of two wheelers and two stroke engine vehicles, increasing congestion, and inadequate and ill-serviced road infrastructure' (Dhakal, 2002: 1).

In response to these air quality issues, the Nepali government started to promote battery-operated electric vehicles and banned the operation of diesel-fuelled three-wheelers, especially in high population density, tourist and heritage places. As early as 1993, a development project, funded by the United States Agency for International Development (USAID) amongst others, assisted the government in starting a pilot project for designing and converting diesel three-wheelers to electric three-wheelers (Plate 5.2). The policies that supported this initiative included a fiscal benefit in the form of reduced import custom tariffs and a waiver on annual vehicle registration fees for electric vehicles. These measures were confirmed in the 2001 National Transport Policy. Moreover, the state-owned Nepal Electricity Authority provided electricity at a low tariff-rate for battery charging. They also introduced off-peak tariffs (night-time charging) to maximize the use of the country's hydropower (Dhakal, 2002).

Kathmandu's experience is a practical example of how measures aimed at improving local problems can address global climate change. More recently, follow-up initiatives for electric vehicles in Kathmandu now explicitly include GHG reduction as a goal, alongside improving air quality. However, Kathmandu also highlights the challenges of mitigation projects. In a 2006 report, USAID (2006: 3) writes:

> The Government of Nepal's policies regarding the electric vehicle sector have been inconsistent, and management of facilities has been poor. Support was needed to encourage formulation of positive regulations (to increase coverage and effectiveness), enforcement standards, and develop other forms of non-polluting transportation, such as bike routes and walking zones. In particular, support was provided to study the feasibility of new approaches and to stimulate the participation of electric vehicle industry associations and public interest groups in a policy dialogue for clean air.

With renewed funding and increased pressure from outside the country, this more recent project achieved a number of policy changes (USAID, 2006: 9). A particular focus was on the tourism industry as a major potential user of electric vehicles:

- removal of registration restrictions on new Safa Tempos (electric three-wheeler) that are used as public transport vehicles (and a redesign for Safa Tempos used

PLATE 5.2 One of about 600 Safa Tempos in Kathmandu. (S. Espiner)

for tourism purposes) – Ministry of Labour and Transport Management and Ministry of Finance;

- reactivation of the Kathmandu Bhaktapur trolley bus services and tax exemption on the importation of trolley bus components – Ministry of Labour and Transport Management, Ministry of Finance and Ministry of Local Development;
- exemption duty increased from 33 per cent to 50 per cent on imported electric- and battery-driven passenger cars and delivery vehicles – Ministry of Finance;
- exemption of excise duty on Safa Tempos – Ministry of Finance; and
- allowing the registration of electric bicycles to operate in Kathmandu – Ministry of Labour and Transport Management.

Key for the future success of transport-related mitigation initiatives in Kathmandu will be the enforcement of existing policies, as well as reducing the dependency of programmes on external sources of funding.

Other examples

Following are a number of additional examples of how tourist destinations have developed policies to address climate change mitigation.

Destination	*Policy*	*Source*
Victoria, Australia	Tourism Victoria's Environmentally Sustainable Tourism Strategic Plan 2009–2012, with details on reducing the industry's carbon footprint (including toolkits).	www.ret.gov.au/tourism/Documents/Tourism%20Policy/workshop%20presentations/Victoria%20Climate%20Change%20DRET%20Presentation.pdf
Great Barrier Reef, Australia	Great Barrier Reef Climate Change Action Plan 2007–2012 and the associated Strategy, includes amongst others an Emissions Calculator and tips for operators to reduce emissions.	www.gbrmpa.org.au
City of Cape Town, South Africa	Cape Town has subscribed to the idea of Responsible Tourism and has developed an extensive programme for its implementation, including environmental management and energy efficiency and a 'responsible tourism pledge' for visitors.	www.capetown.gov.za/en/tourism/Documents/Responsible%20Tourism/Responsible_tourism_bro_web.pdf

5.4 National mitigation policies

Background

Most national policy documents relating to tourism deal with business or management issues. For example, strategic plans or master plans for tourism outline pathways for growth and development and recommend specific policies and actions for marketing, quality control and tourism facilitation. An analysis of the Portuguese PENT (National Strategic Plan for Tourism), and the challenges it poses for climate change mitigation, has been presented by Lobo (2007). Since the plan focuses on strong growth (4.6 per cent annual growth in visitor nights) Lobo concludes that the only way to achieve reductions in GHG emissions from tourism in Portugal lies in an increase in carbon efficiency that outweighs growth rates. Hence the study proposes decoupling the growth of tourism value added from the growth in GHG emissions. This view is shared by the World Travel & Tourism Council (WTTC, 2010).

Tourism departments or ministries are rarely directly involved in climate change policy and seldom give consideration to climate change in tourism policies, as illustrated for Malta and Calvia (Mallorca, Spain) (Dodds and Kelman, 2008). There are a number of reasons, including climate change not being the key mandate of government units responsible for tourism and the lack of resources and expertise required to effectively address climate change mitigation. Some research in this area also provides evidence that tourism stakeholders do not consider climate change to be a major problem, and that communication amongst different government departments (including ones responsible for tourism) is insufficient (Becken and Clapcott, 2011; Dodds and Kelman, 2008). Institutional arrangements in Zambia are noteworthy since there is a Ministry of Tourism, Environment and Natural Resources. The country's climate change unit has been established within this Ministry. Thus the institutional arrangements provide the opportunity for successful climate change adaptation and mitigation in the tourism sector. The Zambian Government is currently considering whether it will develop a specific climate change policy.

Another reason for the general lack of tourism considerations in climate change policy is that in many cases other government departments (e.g. ministries of environment) are responsible for climate change, and they develop climate change policies that apply to all sectors, including tourism (Becken and Clapcott, 2011). In most cases, therefore, tourism is a 'taker' of climate change policy, rather than a 'maker'. A recent OECD and UNEP (2011) report on tourism and climate change confirms this. The review of climate change mitigation policies in OECD member countries highlights that most policies are generic in nature (e.g. related to energy efficiency and transport), and that very few have been developed specifically for tourism. More broadly, there are very few examples of specific environmental policies for tourism, such as environmental taxes (e.g. Aguilo *et al.*, 2005). Against this background, any analysis of climate change policies relevant to tourism might best start by analysing national mitigation policies, as was done by the Pew Centre (now the Center for Climate and Energy Solutions (C2ES)).[3]

Analysis of climate change mitigation policies more generally needs to take into consideration the fact that policies are seldom applied in isolation. In many cases they invoke the use of more than one instrument (see Table 5.2). For example, transportation policies might target GHG emissions but also address problems of traffic congestion or local air pollution. Tourism stakeholders involved in policy making will have to understand these relationships and the implications for tourism.

Case study of a national climate change framework for tourism

The framework 'Australian Tourism and Climate: A Framework for Action' was endorsed by the Australian Tourism Ministers' Council in July 2008. It was developed in partnership with the Australian tourism industry and federal, state and territory governments. Its actions are overseen by the Australian Standing Committee on Tourism (ASCOT) Climate Change Subcommittee. ASCOT more generally develops, coordinates and considers tourism policies in Australia.

TABLE 5.2 Descriptions of selected GHGs abatement policy instruments. (IPCC, 2007b)

Instrument	*Description*
Regulations and Standards	These specify the abatement technologies (technology standard) or minimum requirements for pollution output (performance standard) that are necessary for reducing emissions.
Taxes and Charges	A levy imposed on each unit of undesirable activity by a source.
Tradable Permits	These are also known as marketable permits or cap-and-trade systems. This instrument establishes a limit on aggregate emissions by specified sources, requires each source to hold permits equal to its actual emissions and allows permits to be traded among sources.
Voluntary Agreements (VAs)	An agreement between a government authority and one or more private parties with the aim of achieving environmental objectives or improving environmental performance beyond compliance to regulated obligations. Not all VAs are truly voluntary; some include rewards and/or penalties associated with participating in the agreement or achieving the commitments.
Subsidies and Incentives	Direct payments, tax reductions, price supports or the equivalent thereof from a government to an entity for implementing a practice or performing a specified action.
Information Instruments	Required public disclosure of environmentally related information, generally by industry to consumers. These include labelling programmes and rating and certification systems.
Research and Development (R&D)	Activities that involve direct government funding and investment aimed at generating innovative approaches to mitigation and/or the physical and social infrastructure to reduce emissions. Examples of these are prizes and incentives for technological advances.

The purpose of the climate change framework is to minimize the tourism industry's contribution to GHG emissions, and to prepare the tourism industry to respond to the physical, economic and social impacts of climate change. Most of the framework's actions have been implemented, and outstanding tasks have been transferred to the National Long Term Tourism Strategy's Industry Resilience Working Group. The framework is an excellent example of a holistic approach to tourism and climate change: it addresses both mitigation and adaptation, as well as considering the need for Australia to also address those impacts related to climate change policy in other countries. More specifically the framework sets out five areas of action (Department of Resources, Energy and Tourism, 2008: 5):

1. Improved understanding of the vulnerabilities of tourism to both the physical and economic impacts of climate change in order to build the resilience and

adaptive capacity of the industry and provide certainty for the purpose of future investment.

2. A tourism industry that is prepared for a carbon constrained future and continues to make a substantial contribution to the Australian economy.
3. A repositioning of tourism marketing strategies to meet head on the challenges and opportunities presented by climate change.
4. A fully informed tourism industry through consistent and effective industry outreach and communications.
5. A nationally consistent, inclusive and cooperative approach to implementation.

It is too early to assess the impact of the Australian framework, but at least it can serve as a useful example for other countries that wish to develop a climate change policy framework for tourism. In the meantime, the Australian states have also become involved in a wide range of climate change initiatives (Table 5.3).

TABLE 5.3 Climate change mitigation policies in Australia. (Zeppel and Beaumont, 2011)

	Carbon programme	*Key drivers*	*Carbon tools*
Tourism Queensland	'Sustainability & Climate Change' website; Sustainable Regions Program; Tourism Operator Environmental Indicators Benchmark 2010; Offsetting Your Carbon Footprint fact sheet	Impacts of CC; CPRS; National Carbon Offset Standard; Qld Climate Smart Adaptation; Action Plan; National Action Plan on Tourism & Climate Change	Carbon Reduction Institute; Climate Action Certification; Dept of Climate Change; Earthcheck; ecoBiz; EC3 Global; GHG Protocol; NABERS
Tourism Northern Territory	'Going Green' website; 'Outback Offsets' www.outbackoffsets.com; 'Tourism – Make the Switch' (08/09); Carbon Offsetting fact sheet; Report agency emissions	CPRS; Consumers & CC; Impacts of CC; Five Year Tourism Strategic Plan 2008–2012; NT Govt Climate Change Policy & Coordination Unit; National Action Plan on Tourism & Climate Change	Alice Solar City; Climate Action Certification; Earthcheck; ecoBiz NT; Greenhouse Challenge Plus; NABERS Hotel; PE International (Outback Offsets)

TABLE 5.3 *(Cont'd)*

	Carbon programme	*Key drivers*	*Carbon tools*
Tourism Western Australia	'Climate Change' website; Carbon Footprint Pilot Study (2009)	Consumers & CC; Impacts of CC; WA Office of Climate Change; National Action Plan on Tourism & Climate Change; National Climate Change Adaptation Framework; Climate Change Guide	Carbon Neutral; Dept of Climate Change; DEWHA GHG Protocol; ISO Standard Sustainable Infrastructure Australia (SIA); WA Small Business Development Corporation
Tourism Victoria	'Tourism Excellence' website; Tourism destination carbon footprint study; Audit own carbon footprint	Impacts of CC; Consumers & CC; National Climate Change Adaptation Framework; National Action Plan on Tourism & Climate Change; CPRS; Our Environment, Our Future (Vic. Govt Strategy)	CarbonDown; Carbon Innovators Network; Climate Action Certification; Greenhouse Challenge Plus; Grow Me The Money; Sustainability Victoria
Tourism Tasmania	'Climate Change' website; Green Tourism www.greentourism.com.au	Consumers & CC; National Tourism & Climate; Change Taskforce; Tas. Climate Change Strategy; Reducing Tasmanian Government's Carbon Footprint; CPRS	Climate Change Office Tas.; Climate Managers; EarthcheckEC3 Global; EcoSmartBiz; Greenhouse Challenge Plus; Tony Charters & Associates
South Australia Tourism Commission	Green SA; Environmentally Sound Technology; Resources	CPRS & higher costs; Promote low carbon tourism	Climate Action Certification

Notes: CC = Climate change; CPRS = Carbon Pollution Reduction Scheme; DEWHA = Department of Environment, Water, Heritage & the Arts; GHG = Greenhouse gas; ISO = International Standards Organisation; NABERS = National Australian Built Environment Rating System

Case study of an aviation duty

The British Air Passenger Departure Duty (APD) is the most prominent example of a tax or duty on air travel. It has generated a storm of protest amongst airlines, tourism businesses and destinations, all of which claim that the ongoing increases in duty severely undermine outbound tourism from the United Kingdom (Plate 5.3).

More specifically, the APD is an excise duty which is levied on the carriage, from a UK airport, of chargeable passengers on chargeable aircraft. APD came into effect on 1 November 1994 with the goal to simply raise revenue for the Government. The tax has been increased several times (HM Revenue and Customs, 2010). From 1 November 2009 a four-destination band structure based on geographical distance from London was applied. The rates, which also depend upon the class of travel, increased again in 2010 (Table 5.4).

While the British Government has explicitly stated that the APD is not an environmental charge, it has been interpreted by some as providing 'green signals'. The Government itself has argued that the climate change cost of aviation will be covered by the EU Emissions Trading Scheme charges from 2012. However, for those who see the APD as a 'climate tax' (e.g. the tourism industry and IATA) this results in double dipping. The APD has also been interpreted as a climate change instrument by environmental advocacy groups (e.g. the Worldwide Fund for Nature (WWF) UK) who welcome it as part of their campaigning against air travel and airport extensions. The confusion about what the tax is designed for even extends to government agencies. A 2008 study by the Department of Transport examined whether the APD would fully cover the costs of climate change under

PLATE 5.3 Passengers at Gatwick Airport, London. It is the UK's second largest airport and the busiest single-runway airport in the world. It serves over 200 destinations for around 33 million passengers a year. (S. Becken)

TABLE 5.4 APD rates of duty from 1 November 2009. (HM Revenue and Customs, 2010)

		From 1 November 2009 to 31 October 2010 (£)	*From 1 November 2010 (£)*
Standard rates	Band A (0–2,000 miles)	22	24
	Band B (2,001–4,000 miles)	90	120
	Band C (4,001–6,000 miles)	100	150
	Band D (over 6,000 miles)	110	170
Reduced rates (elaborate)	Band A (0–2,000 miles)	11	12
	Band B (2,001–4,000 miles)	45	60
	Band C (4,001–6,000 miles)	50	75
	Band D (over 6,000 miles)	55	85

a range of assumed shadow prices for carbon. The study found that under the baseline scenario, and the past levels of duty, the external cost of climate change was not covered, but that the anticipated increases in duty (as effective now) would cover the carbon costs. This implies that the APD is indeed a climate tax.

In summary, it remains unclear whether the APD is a climate change instrument or not. Notwithstanding this, it provides a useful example of what an aviation policy could look like and what impacts it might have on tourism. Germany has recently introduced a similar departure tax for aviation, but doubts have been raised about how long it might last based on experiences in other European countries that introduced aviation taxes and then abolished them after strong resistance (e.g. the Netherlands) (Gordijn, 2010).

Case study of a national energy plan with a focus on tourism

In Hawai'i, United States, the government department that deals with matters related to tourism (as part of economic development) is also mandated with energy policy making. Thus Hawai'i's Department of Business, Economic Development and Tourism is in a strong position to develop policies that are relevant to climate change and align with the goals of tourism development.

Hawai'i is the most oil-dependent state in the United States, relying on imported petroleum for about 90 per cent of its primary energy requirements. Energy costs are therefore high. To decrease dependency, and reduce costs for businesses and households, the Government has developed a wide range of policies and programmes.[4] While reduction in GHG emissions is not the key driver of these policies, it is a major beneficiary. Relevant policies relate to:

- commercial and government buildings;
- educational buildings;

- the Building Energy Code;
- mandatory solar water heaters for all new single-family dwellings;
- the Refrigerator Rebate Program; and
- the Green Business Program.

The Green Business Program is particularly relevant as it initially focused on the hotel industry through the Hotel Industry Program partnership. The programme is designed to help businesses go beyond compliance to conserve energy, water and other resources, and to reduce pollution and waste. This programme also acts as an information-sharing platform. The following tourism organizations are involved: Hawaii Hotel and Lodging Association; Hilton Hawaiian Village Beach Resort and Spa; and Sheraton Hotels and Resorts in Hawaii. The lessons learned from the hotel industry will be used in future initiatives for offices and commercial buildings.

Other examples of national climate change mitigation policy for tourism

Several countries have become proactive in addressing the issue of climate change and tourism at a policy level. These range from 'mainstreaming' tourism into a national climate change mitigation strategy, as in the case of Austria, to specifically developing plans or strategies for the tourism sector (e.g. in New Zealand).

Country	*Policy*	*Source*
Austria	The 2008–2012 Climate Change Mitigation Strategy by the Lebensministerium provides a good example of how tourism is integrated as a vital sector and activity (e.g. in mobility management and energy efficiency in buildings) into a national plan.	http://umwelt.lebensministerium.at/article/articleview/70701/1/7337
New Zealand	Industry and government have worked together to develop a Tourism and Climate Change Plan which identifies the key priorities and highlights actions that can be taken to help the sector respond effectively. Mitigation is an integral part of the plan.	http://www.tourism.govt.nz/Our-Work/Our-Work-Summary-page/Climate-Change/

Country	*Policy*	*Source*
Barbados	Government policy measures that encouraged diffusion of solar water heating with a particular focus on tourism: • Waiving of taxes on raw materials for solar water heater manufacture; • High taxes on non-solar water heaters (>60%); • Hotels that borrowed from the government-run development bank had to carry out energy audits.	Yaw (2005)
Ireland	Fáilte Ireland developed a Carbon Strategy (2008) that details seven key actions to be implemented in partnership with various industry stakeholders, in particular Tourism Ireland and the Irish Tourism Industry Confederation.	http://www.failteireland.ie/Information-Centre/Publications/Environment/Carbon_Strategy_Brochure_2009
Germany	From the 1st of January 2011 there has been a departure tax on every flight leaving a German airport. Flights under 2,500 km will cost 8 Euro, those over 6,000 km 25 Euro and 5 Euro for all other flights.	http://www.bund.net/fileadmin/bundnet/pdfs/bundletter/20101109_verkehr_flugverkehrssteuer_hintergrund.pdf

5.5 Regional level mitigation policy

Background

Several regions are active in addressing GHG emissions from tourism, for example the Caribbean, the South Pacific and the European Union. The regional level is important as it provides an opportunity to join forces and pool resources, and to develop coordinated action that is likely to be more effective than initiatives by an individual country. The regional approach is particularly important in the case of many small countries that wish to gain a greater collective weight and also be more prominent on a global stage (such as the Alliance of Small Island States). In the following sections, policies by the European Union will be discussed from a tourism perspective, followed by a case study on the Caribbean.

Case study: The European Union

Tourism is not an explicit policy area for the European Union (EU). However, in October 2010, the EU indicated that it will develop a policy on tourism. This new policy would explicitly address climate change, whereby any actions in support of tourism should be carried out on the basis of the four axes outlined by the European Commission:

1. Stimulating competitiveness in the European tourism sector.
2. Promoting the development of sustainable, responsible and high-quality tourism.
3. Consolidating the image and profile of Europe as a collection of sustainable and high-quality destinations.
4. Maximizing the potential of EU policies and existing financial instruments for developing tourism.

Further, the need to improve the knowledge base on climate change was recognized in order to be able to provide useful information to the tourism industry's strategies, and the policies of public authorities.[5] The EU can build on the extensive policies already in place in the area of energy (Figure 5.5), environment and transport. The 2008 policy package on energy and climate change, for example, was introduced with the aim of making Europe the world leader in renewable energy and low carbon technologies.

The most widely debated EU policy on climate change in relation to tourism is the Emissions Trading Scheme (EU ETS) and the imminent inclusion of aviation emissions. The EU ETS was established in 2003 (Directive 2003/87/EC) and included land-based industrial installations. However, from 1 January 2012 flights arriving at, and departing from, European airports will be part of the EU ETS. Aircraft operators have already started fulfilling their obligations by submitting their monitoring plans to competent authorities in 2009 and monitoring their emissions and tonne per kilometre data in 2010.[6]

In July 2011, the European Commission released a baseline figure for the historical emissions from aviation in the European Economic Area (EEA). Calculated as the average of the emissions in 2004, 2005 and 2006, the CO_2 emissions amounted to 221,420,279 tonnes per annum. The quantity of emissions allowed for flights in and out of the EEA will be equal to this amount initially, but are reducing to 97 per cent in 2012 and 95 per cent in 2013–2020. Of the emission allowances in 2012, 85 per cent will be allocated to current operators and 15 per cent will be auctioned. The freely allocated allowances will be reduced to 82 per cent for the 2013–2020 period, and 3 per cent will be saved as special reserves (e.g. for new airlines entering

POLICY DOCUMENTS FOR ENERGY

- European Energy Programme for Recovery
- Energy Security and Solidarity Action Plan
- An Energy Policy for Europe
- Green Paper. A European strategy for sustainable, competitive and secure energy
- Factors contributing to the development of energy policy
  - Information on investment projects in energy infrastructure
  - 'Green' vehicles: a European strategy
  - Reducing greenhouse gases by 2020
  - ICTs to facilitate the transition to an energy-efficient, low carbon economy
  - Strategy on climate change: the way ahead for 2020 and beyond
  - Strategy for sustainable development
  - Integrating the environment into community energy policy
  - European Energy and Transport Forum

MARKET-BASED INSTRUMENTS

- Market-based instruments for the environment
- Greenhouse gas emission allowance trading scheme
- Community framework for the taxation of energy products and electricity

RESEARCH AND INNOVATION

- SET-Plan for the development of low carbon technologies
- Sustainable power generation from fossil fuels
- Demonstration of the capture and storage of CO_2
- Strategic Energy Technology Plan (SET Plan)
- Towards a European Strategic Energy Technology Plan

FINANCIAL INSTRUMENTS

- Competitiveness and Innovation Framework Programme (CIP) (2007–2013)
- Seventh Framework Programme (2007–2013)

FIGURE 5.5 Energy policies of the European Union. (After http://europa.eu/legislation_summaries/energy/european_energy_policy/index_en.htm)

the market). Airlines are allowed to use Certified Emission Reductions (CERs) and Emission Reduction Units (ERUs) up to 15 per cent of their EU ETS allocation in 2012, but from 2013 the usage of CERs remains unclear (Anger and Koehler, 2010).

The inclusion of aviation into the EU ETS has been heavily opposed by many countries outside the European Union, including the United States and New Zealand, as it is seen as a major barrier to trade and tourism. The example of the EU is informative as it highlights how a region – in the absence of a global agreement – can make progress in climate change mitigation, and by doing so potentially act as a catalyst and foundation for a comprehensive and inclusive international agreement. This is based on the assumption that the aviation component of the ETS will work and that the targets are sufficiently ambitious.

Airlines will be included in EU ETS from 1 January 2012 but, for many of us, building the cost of carbon into our major business decisions is nothing new. To really deliver on climate change targets future policy frameworks for carbon must be both environmentally effective and economically efficient, and for aviation this means working at the global level. But it will also need businesses and global policy makers to work together creatively, to bring consumers along with us. (Steve Ridgway, Chief Executive, Virgin Atlantic)

Case study: The Caribbean

Over recent years the Caribbean has been very active in relation to climate change. At a higher level, the Caribbean Renewable Energy Development Programme (CREDP) focuses on both the reduction of the Caribbean region's dependency on fossil fuels and a reduction in GHG emissions (Simpson *et al.*, 2008b). Given that in many Caribbean countries hotels account for more than half of national energy use this programme is very important for the tourism sector, as it aims to create a framework for facilitating renewable energy investments in selected Caribbean countries. The most important energy sources will be solar heating for hot water systems, photovoltaic electricity generation, hydropower and biofuels (Plate 5.4).

PLATE 5.4 Solar lamp, lighting the garden environments in a hotel in Tobago. (R. Mahon)

The Caribbean Hotel Energy Efficiency Action Program (CHENACT) (Duffy-Mayers, 2010) is a more recent initiative specifically designed to support hotels. The region's public- and private-sector tourism organizations, the Caribbean Tourism Organization (CTO) and the Caribbean Hotel and Tourism Association (CHTA) – the Caribbean Alliance for Sustainable Tourism (CAST) – started a US$2 million project in 2009 to help the Caribbean hotel sector reduce energy consumption.[7] The beneficiaries (directly or indirectly) will be the 2,670 hotels in 25 countries and territories of the Caribbean.

The objective is to improve the competitiveness of small- and medium-sized hotels (<400 rooms) through improved energy use, with an emphasis on renewable energy generation and micro-generation. The pilot project is in Barbados and entails detailed energy audits and analyses of energy consumption patterns among Caribbean hotels. The programme is currently developing a Clean Energy Policy for the hotel sector in Barbados. This will be replicated throughout the Caribbean. The policy needs to take into account that energy efficiency and the introduction of renewable energy generation have a wide range of effects, some of which might

TABLE 5.5 Impacts of the suggested energy policies for Barbados, based on detailed energy audits and analysis. (Duffy-Mayers, 2010)

Wider impacts from potential energy efficient savings	*Impacts from potential solar PV renewable generation in the hotel sector*
• Equivalent barrels of oil saved in electricity generation: 49,200 barrels per year • Potential reduction of greenhouse gas emissions: 28,600 CO_2 tons per year • Financial impact for the power utility (BLPC) – US$8.2 million reduced revenue per year as result of reduced sales • Fiscal impact for the government (without including grants): – Loss of US$1.2 million per year from VAT revenues over utility's electricity sales – Loss of US$800,000 in customs duty revenues, due to 5% customs duty incentive for EE equipment – Loss of US$600,000 per year in income tax revenues due to the 150% write-off of Loan Interest Income Tax concession – US$7 million in Income Tax Credits as included in the Tourism Development Act – Reduction of US$3.7 million per year in the annual imports – exports deficit due to fewer oil imports – Increase of approximately US$1 million revenues from VAT on EE goods and services	• Equivalent barrels of oil saved in electricity generation: 3,200 barrels per year • Potential reduction of greenhouse gas emissions: 1,875 CO_2 tons per year • Financial impact for the power utility (BLPC) – US$550,000 per year as result of reduced sales • Fiscal impact for the government (without including grants): – Loss of US$80,000 per year from VAT revenues over utility's electricity sales – Reduction of US$200,000 per year in the annual imports of oil

be unanticipated or undesirable (Table 5.5). For example, the reduction in electricity consumption will result in decreased revenue for government as a result of less value-added tax collected.

Based on the Barbados experiences, the CHENACT programme further advocates use of the following policy instruments and interventions for Caribbean countries (Duffy-Mayers, 2010):

- Tourism Development Act – to establish a preference for energy efficiency versus non-energy efficiency investments and increase incentives for solar PV;
- Renewable energy system interconnection pilot programme – a standardized long-term power-purchase agreement.
- Utility Demand Side Management (DSM) programmes – to encourage the recovery of utility expenditures and allow a return on investment through the rate base.
- Building code (draft energy efficiency code) – applied to lighting, air-conditioning, hot water and appliance standards for hotel applications.
- SMART Energy Fund Pre-investment grants – low-interest loans tied to payback period, performance risk sharing and upfront payments for Certified Emission Reductions (CERs).
- GHG Mitigation Action Plan – business-as-usual forecast for hotel sector, emissions reduction targets, registration of clean development mechanism projects and emissions reduction monitoring and verification.
- Carbon Neutrality Programme – promotion and marketing of 'low carbon' development in tourism sector and accreditation of 'low carbon' hotel properties.

The Caribbean example of an energy efficiency programme for hotels highlights that there can be a significant multiplier effect if projects use limited resources in a way that – while piloted in a particular location or context – are applicable to a wider region that is characterized by common parameters for implementation.

Other examples of mitigation policies at a regional level

Regional level initiatives in the field of climate change mitigation and tourism policy are not widespread, but five more examples are presented below.

Region	*Policy*	*Source*
South Pacific Islands	The South Pacific Tourism Organisation developed a strategic plan for 'Green Tourism' with specific consideration of climate change and carbon issues.	www.south-pacific.travel/spto/export/sites/spto/tourism_resources/green_tourism/draft_final_green_ras_31_oct.pdf

Region	*Policy*	*Source*
Mediterranean	The Greek Government, with the support of the European Investment Bank (EIB) and in collaboration with leaders from across the Mediterranean, launched the Mediterranean Climate Change Initiative on 22 October 2010.	http://climate-l.iisd.org/news/mediterranean-climate-change-initiative-launched/
Arab Climate Resilience Initiative	Focuses on Middle East and North Africa countries (MENA) and funded by UNEP to address both mitigation and adaptation.	www.arabclimateinitiative.org/
EU Hotel Energy Solutions Project, Europe	Public-private sector partnership, supported by UNWTO and UNEP. The goal is to increase energy efficiency in the European hotel sector by 20% and the use of renewable energy sources by 10%.	Hotel Energy Solutions www.hotelenergysolutions.net/
Insider Guide to Climate Change, Great Britain	Guide for tourism managers to reduce carbon emissions.	www.thetravelfoundation.org.uk/assets/files/our_work_with_industry/climate_change/insider_guide/CLIMATE%20INSIDER%20GUIDE%202010.pdf

5.6 International mitigation policy for tourism

Background

The most prominent international agreements related to climate change mitigation are the United Nations Framework Convention on Climate Change (UNFCCC) and Kyoto Protocol. While only the latter is specific to mitigation, both have been instrumental in stimulating a common approach towards climate change mitigation. However, neither has resulted in deep cuts in emissions, frustrating environmental organizations and climate activists alike (e.g. Greenpeace, Plate 5.5). The IPCC (2007a) notes that any future global agreement will need to cover the majority of emissions, including those from emerging developing countries. To limit the temperature increase to 2°C above pre-industrial levels, developed countries would need to reduce emissions in 2020 by 10–40 per cent below 1990 levels and in 2050 by approximately 40–95 per cent. Further, developing countries would need to change their emission path by 2020.

The United Nations Climate Change Conference in Cancun (2010) was hosted by the Government of Mexico. It achieved some modest progress, most notably the Cancun Agreement. The agreement represents a ‘key step forward in capturing plans

PLATE 5.5 Greenpeace at the 2007 UNFCCC Climate Summit in Bali, Indonesia. (S. Becken)

to reduce greenhouse gas emissions and to help developing nations protect themselves from climate impacts and build their own sustainable futures' (UNFCCC, 2011). Three major elements are noteworthy. First, the Cancun Agreement is a formal, collective effort to reduce GHG emissions, under the banner of the UNFCCC. Second, it contains a comprehensive package to help developing nations address climate change, in terms of both mitigation and adaptation. Third, the agreement contains a schedule for nations to review the progress they make towards their expressed climate change objectives.

Only recently has tourism had some level of involvement in international climate policy, albeit at a very limited scale and not to the extent of achieving any international agreement. The Djerba Declaration on Climate Change and Tourism developed by UNWTO, along with several other United Nations agencies, was followed by the Davos Declaration on Tourism and Climate Change in 2007 (Scott and Becken, 2010). This declaration was also endorsed by governments at the Tourism Ministers Summit on Climate Change (London 2007) and presented at the Copenhagen Climate Summit in 2009. Since those first important steps, the UNWTO position on climate change has been called 'fuzzy' (Institute for Integrative

Tourism & Development, 2009: 11). Within the framework of the UNFCCC the UN agency for aviation, the International Civil Aviation Organisation (ICAO), is facilitating ongoing negotiations in relation to an international agreement on aviation emissions but with little success to date. Cross-industry and public–private sector partnerships, as well as political commitment, will be key to address the climate change challenges for tourism.

> Together, we must put in place the frameworks for implementing the technology and infrastructure needed to decarbonise air transport. … For tourism there needs to be more collaboration across the value chain on creative innovations, which can allow the whole sector to 'decarbonise'. (South Africa Tourism Minister Van Schlakwyk at the World Economic Forum (WEF) Africa Pre-summit on Green and Shared Growth in Aviation, Travel and Tourism in Cape Town, 4 May 2011)

The following two case studies discuss the efforts of two global industry bodies, one for the travel and tourism industry and one relating to the aviation sector. Each of these, it appears, has been proactive in developing mitigation policy in the absence of tangible efforts by UN agencies formally representing these sectors (i.e. the UNWTO and ICAO).

Case study of an international industry association

The World Travel & Tourism Council represents the world's leading travel and tourism companies. One of its main tasks is to raise awareness of tourism as a globally important industry, and to foster sustainable practices in the sector. The WTTC divides its activities into global initiatives (e.g. the organization of the Global Travel and Tourism Summit), regional initiatives and economic research. In the last few years the WTTC has become increasingly active in the area of climate change, for example through its participation at the Copenhagen Climate Summit in 2009. 'Climate Change – A Joint Approach to Addressing Climate Change', prepared by the WTTC, details its priorities, commitments and suggestions for cooperation with government agencies in relation to climate change mitigation (WTTC, 2010).

In this document, the WTTC (including 40 chief executives of leading tourism companies) announced the sector's goals for CO_2 emission reductions and acknowledged that the 'recent model of carbon-intensive tourism growth, consumption and production is now under review' (WTTC, 2010: 3). They suggested that tourism can and should take a leadership role in climate change mitigation, especially in those countries where tourism dominates the national economy. The proposed areas of policy and action are organized under five themes (see below excerpt from document), with action items for each. In addition, the document provides more detailed information on specific industry initiatives and suggestions for government actions within each of the five themes. These could serve as a useful guideline for national governments in their development of climate policy for tourism.

Theme 1: Accountability and responsibility

- Aspire to reduce CO_2 emissions by 50 per cent by 2035 (with an interim target of a CO_2 emissions reduction of 30 per cent by 2020 if there is an international agreement on global emissions reduction, or by 25 per cent by 2020 in the absence of such an agreement).
- Align efforts to produce an internationally agreed framework of standards to measure progress against GHG emission targets.

Theme 2: Local community, sustainable growth and capacity building in climate change hot spots

- Identify climate change hot spots in which we operate, develop practical strategies and create examples of low carbon businesses, helping local communities access information on new technologies to reduce their own emissions.
- Support efforts to reduce emissions from deforestation and forest degradation and develop links with partners including the Prince of Wales Rainforest Project.

Theme 3: Educating customers and stakeholders

- Promote incentives to encourage customers to play a significant part in reducing the use of energy and lead by example with our sustainable business practice.
- Devote resources to explaining our climate change strategies and share our experience of exploring new low carbon technologies and sustainable approaches that mitigate GHG emissions.

Theme 4: Greening supply chains

- Develop a range of practical tools that can be used by small and medium enterprises (SMEs) to measure and share their CO_2 emission levels.

Theme 5: Innovation, capital investment and infrastructure

- Endorse and support the latest progressive CO_2 emissions' reduction commitments from our Members in the aviation sector and encourage all airlines to commit to achieving at least a 50 per cent reduction in emissions by 2050.

- Adopt environmental management systems to continually measure our energy use and GHG emissions, design an e-platform of best sustainability practices per sector (e.g. energy, water, solar, waste, etc.) that engage with our staff and customers and share our progress towards meeting annual energy and emissions' reduction targets with all stakeholders.
- Target the investment community and encourage the creation of financial instruments closely linked to sustainable projects in the Travel & Tourism industry.

While representing a useful beginning, the WTTC approach has also been criticized by tourism academics because of the targets being merely 'aspirational' and potentially even unrealistic (Weaver, 2011). The targets are more ambitious than any other tourism targets (e.g. from the aviation sector, see below) and it is hard to see how tourism as a whole can achieve substantial cuts without major players on board (Institute for Integrative Tourism and Development, 2009). Some industry players (see quote by Ryanair Chief Executive Michael O'Leary below) are not supporting environmental approaches in the tourism and travel industry, including aviation.

> We will double our emissions in the next five years because we are doubling our traffic. But if preserving the environment means stopping poor people flying so only the rich can fly, then screw it. (Michael O'Leary, Chief Executive of Ryanair)

Case study of the airline industry

The International Air Transport Association (IATA) provides global statistics, analysis and policy related to aviation. Most recently, for example, the IATA reports 'robust recovery' of the aviation sector in 2010, with a growth of 6.3 per cent in passenger numbers compared with 2009 (IATA, 2011). IATA's advocacy for the industry has also led it to take a proactive role in climate change mitigation policy, partly in response to negative media coverage and the risk of air travel being identified as an environmentally irresponsible activity. Figure 5.6 shows the anticipated pathway of CO_2 emissions under a business-as-usual scenario (i.e. growth as forecast), and reductions due to technological improvements, air traffic management and alternative fuels.

Officially, IATA's climate change response hinges on four pillars: (i) technology; (ii) operations; (iii) infrastructure; and (iv) economic measures (IATA, 2009a) (Figure 5.7). Reductions due to economic measures that might impact on visitor demand are interestingly not depicted in Figure 5.6. More technical details on technological reduction options are provided in Chapter 6 (Mitigation Practice).

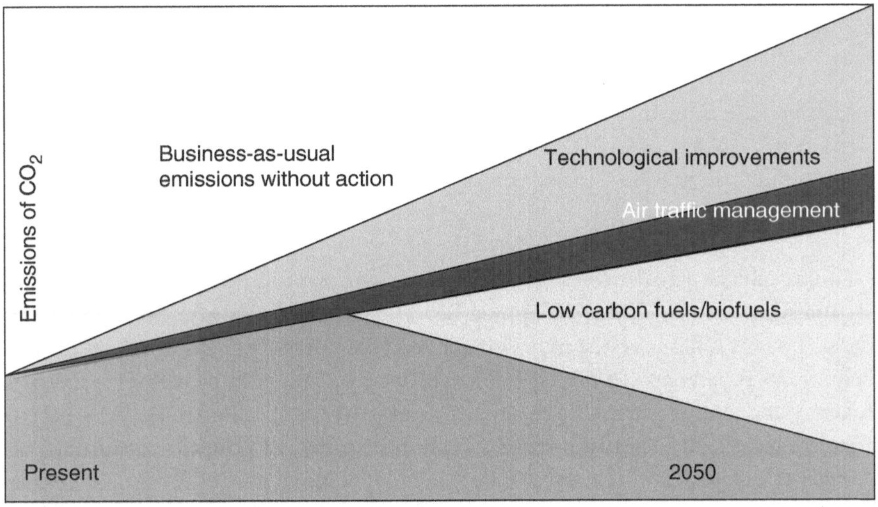

FIGURE 5.6 IATA Conceptual Emission Reduction Scenario. (After Steele, 2009)

Pillar 1: Technology
Short Term: Incorporation of Service Bulletins that reduce fuel burn and aircraft weight;
Medium Term: Airline fleet renewal with new aircraft and new engine technologies;
Longer Term: Entire new aircraft design;
IATA's target is for 10% of the fuel used by aircraft to be an alternative fuel by 2017.

Pillar 2: Operations
In coordination with air traffic management improvements, operational efficiency improvements can make a big difference in fuel savings;
More efficient flight planning can reduce fuel reserve requirements;
Aggressive weight reduction programmes save fuel;
Further improvements will be realised with upgrades to the avionics system on board airplanes coupled with an improved air traffic control system.

Pillar 3: Infrastructure
The technical elements of air traffic services, such as communication, navigation, surveillance, separation minima and air traffic flow management that have a direct impact onto the availability of optimum flight profiles in the terms of speed, route of flight, climb/descent profile and altitude;
Airport infrastructure development and the elements involved with movement of passengers, aircraft movement and servicing landed aircraft can also play a role in saving emissions;
Many of the infrastructure challenges are not technical, and IATA works with governments and authorities to help achieve optimum solutions.

Pillar 4: Economic Measures
Positive economic instruments to provide incentives to improve efficiency and reduce emissions;
Emissions trading and carbon offsets can play a role but this needs global coordination and agreement. Credible global standards are essential.

FIGURE 5.7 The four pillars to climate change mitigation by IATA. (After IATA, 2009a)

In 2007 the four-pillar strategy was adopted by the global aviation industry, as well as by International Civil Aviation Organisation (ICAO) member states, while in 2009 IATA airlines adopted a set of ambitious targets:

1. A cap on aviation CO_2 emissions from 2020 (carbon-neutral growth).
2. An average improvement in fuel efficiency of 1.5 per cent per year from 2009 to 2020.
3. A reduction in CO_2 emissions of 50 per cent by 2050, relative to 2005 levels

These collective goals were endorsed by the aviation industry in the joint industry submission to the ICAO in September 2009 (IATA, 2009a). To support the targets, the IATA has invested in a wide range of studies and reports, for example the Technology Roadmap (IATA, 2009b) that details possible improvements for fuel efficiency and alternative fuels (see also Chapter 6, Mitigation Practice). While these initiatives have not been supported by technical studies of how the cuts might be achieved (Institute for Integrative Tourism & Development, 2009), the IATA has been bolder in its climate change communication (or public relations activities) than the ICAO, even though the latter is officially mandated to address aviation's climate impacts. Whether the IATA and its members will be able to achieve their ambitious goals will only be seen over time. Even now, some experts in the field question the feasibility of the proposed targets (e.g. Peeters and Middel, 2007; Scott *et al.*, 2010).

Other examples

Many of the international tourism organizations have dedicated specific initiatives or policies to climate change mitigation. Some, like the Aviation Global Deal Group, have been formed specifically to influence climate change policy.

Organisation	*Policy*	*Source*
The International Ecotourism Society	Sustainable Transportation Guidelines for Nature-based Tour Operators (with details on climate change mitigation).	www.ecotourism.org/site/c.orLQKXPCLmF/b.4835379/k.55C1/TIES_Publications__The_International_Ecotourism_Society.htm
Travelife, Europe	Collaboration of European tour operators and associations working on a common approach in the industry towards sustainability.	www.travelife.eu

Organisation	*Policy*	*Source*
Sustainable Tourism Alliance	USAID programme that allows tourism destinations in developing countries to access world-class expertise with proven success in addressing sustainable development challenges, including climate change.	www.gstalliance.net
Aviation Global Deal Group	The Aviation Global Deal (AGD) Group is an industry coalition of leading international airlines, aviation sector companies and the international NGO The Climate Group. Their goal is to contribute towards a pragmatic, fair and effective policy solution that incorporates international aviation CO_2 emissions into a global climate change deal.	www.agdgroup.org

5.7 Conclusion

In the last five years tourism has become more involved in policies relating to climate change at many levels. The case studies of tourism businesses show that tourism company policies, which include climate change mitigation, typically build on wider environmental management initiatives, with reducing GHG emissions being added as an explicit goal alongside others. There are many examples across the international tourism industry and therefore selecting representative case studies is rather challenging.

Policy making at the public sector or industry association level is slightly more complex. At these levels the tourism sector can be involved in policy making relevant to climate change in four distinct ways (Becken and Clapcott, 2011):

1. Passively: tourism interests are sufficiently and equitably covered within national climate change policy (e.g. a generic carbon tax).
2. Passively: tourism is sufficiently and equitably considered in national policy that addresses issues indirectly related to climate change (e.g. biodiversity protection, air pollution) – no specific action is required.
3. Advocacy: tourism advocates the specific inclusion of climate change aspects into national policies that address issues that are important to tourism, but that fail to take sufficient account of climate change (e.g. water-related legislation).

4. Actively: tourism develops its own 'climate change policies', specifically for the tourism sector (e.g. an environmental quality scheme that includes carbon emissions).

This chapter has provided examples for each of the above types of involvement, at both local (i.e. destination) and national levels. This chapter also discussed policy initiatives that are occurring at a regional and international level.

The case studies of policies which deal with climate change mitigation actively developed by (rather than for) the tourism sector support the view that these are largely designed to protect the sector, rather than reduce the risk of climate change per se (Becken and Clapcott, 2011). Thus, tourism's policy making in the climate change space is likely to be 'tourism policy' rather than 'climate policy'. This is possibly explained by tourism's propensity to prefer approaches of marketing and communication, rather than management. It may also reflect the institutional arrangements where tourism departments or organizations are mandated to undertake marketing activities, thus playing an advocacy rather than a regulatory role. Another feature is that most policies prepared by tourism organizations are voluntary in nature. While this may be a pragmatic starting point for tourism to become more involved in climate change initiatives, to be effective in the long term these need to be embedded in a broader policy and regulatory mix (Krarup and Ramesohl, 2002).

If the tourism sector is to play a major role in national and international climate change policy making, stronger leadership by key organisations such as UNWTO and ICAO is essential. At present all organisations indicate they want to be somehow involved in the climate debate (or cannot afford to be marginalised), but real commitments are lacking. There is a tendency to 'wait and see', with short-term economic goals outweighing longer-term requirements to manage climate change risks.

6

MITIGATION PRACTICE

> If you think mitigated climate change is expensive, try unmitigated climate change.
>
> *(Prof. Richard Gammon, Harvard University)*

6.1 Introduction

Tourism entails many different activities and services. These include, amongst others, marketing activities and the provision of transport services, accommodation, and recreational activities. Tourist attractions or activities are often provided by a commercial operator (e.g. a boat trip), but can also include publicly managed attractions such as national parks or museums. Tourism therefore cuts across the private sector (i.e. the businesses), the public sector (i.e. government) and communities (see also Becken and Hay, 2007). Accordingly, both the range of activities that constitute tourism and the number of stakeholders are large. Table 6.1 provides an overview of possible mitigation measures across different tourism activities and stakeholders. The table provides examples of the initiatives that destination planners, tour operators and tourists can take to mitigate GHG emissions across a range of sub-sectors.

The remainder of this chapter provides detailed case studies and examples of the wide range of mitigation opportunities in tourism. The first section, on destination appeal, highlights how mitigation is either driven by the local government and its community (e.g. in the case of Växjö) or directly by the tourism sector, as in the case of the Seychelles or Kangaroo Valley. Many of the tourism-driven initiatives are motivated by a desire to improve or maintain a favourable image of an environmentally friendly destination, especially in the case of nature or 'wellness' tourism.

TABLE 6.1 Examples of mitigation measures for tourism. (After Simpson *et al.*, 2008b)

Stakeholder	*Destination marketing*	*Aviation*	*Land transport*	*Accommodation*	*Attractions*
Destination managers/ planners (e.g. councils)	Work closely with marketing organizations to align brand (e.g. 'green') with wider destination and community activities	Encourage carbon efficient airlines; focus on revenue and yield not growth	Encourage provision and use of public transport and small rental cars; provide infrastructure for walking and cycling	Encourage local businesses to engage in mitigation; provide support and expertise (e.g. energy consultants)	Encourage local businesses to engage in mitigation; provide support and expertise (e.g. facilitate information exchange)
Tour operator	Offer close destinations; provide carbon labelling either by working with existing ecolabel organizations or by carbon footprinting products	Book clients on airlines that mitigate their emissions; select most energy efficient routes	Include train or bus in packages; book fuel efficient vehicles	Recommend energy/ environment certified hotels	Offer to include low carbon activities in package
Tourist	Stay longer; visit closer destinations (e.g. domestic holiday)	Minimize air transport; choose climate-friendly airline; offset emissions	Use train and coach, rather than plane or car; use energy efficient cars (<120 g CO_2/km)	Choose environmentally certified hotel; switch off lights and air conditioning when leaving room	Reduce energy intense activities; choose providers that are energy/ environment certified

The section on transport provides a broad overview of the carbon intensity of different transport modes and options for mitigation. It then goes on to discuss in more detail the achievements of the aviation sector to date, and what it is aiming to do in the future to reduce GHG emissions. Aviation efforts typically rest on three types of initiative, namely technological improvements, air traffic management and alternative, more sustainable, fuels. Transport case studies discussed in more detail also include bus transport, bicycle tourism as a form of 'slow travel', and innovation on board a small cruise ship. These three examples highlight a great potential for mitigation, although – at present – these modes represent niche markets in the overall picture of tourism transport.

Tourism facilities such as hotels and conference centres are key targets for energy savings and climate change mitigation. The section on facilities and operations uses three case studies to demonstrate achievements and opportunities. The first case study relates to a self-contained accommodation business in the United Kingdom. In this case innovative measures have been combined with standard activities, such as changing to energy efficient light bulbs. Collectively they have led to substantial decreases in energy consumption. These learning experiences are, in turn, used to educate guests.

The second case study is based on the ski resort of Whistler Blackcomb. This case study highlights the range of options for operators of tourist activities to mitigate their climate impact. Finally, the case study relating to a tour operator is used to illustrate cross-cutting mitigation options for those who 'bundle' activities into a package of services that target tourists. A large number of other examples is provided in a concluding table.

The final section in this chapter discusses the role that tourists can play in climate change mitigation. Increasingly tourists are informed about, and aware of, the consequences of their travel for the climate. While this has not yet been translated into widespread behavioural changes, there are examples where tourists are actively engaging in mitigation. One such case that is discussed in more detail relates to the provision of climate information, such as the carbon content of particular food products or package tours. Another example relates to carbon offsetting, which is popular amongst a small, but growing, segment of tourists. The German Atmosfair programme is used to illustrate how carbon offsetting works for tourism.

6.2 Destination appeal

Background

Tourist destinations are locations and regions in which tourism plays a key role in the local economy. They can be an individual city, a district, a national park, or any other geographic unit, such as the Mekong Delta in Southeast Asia. The latter is a tourist destination which includes several countries. Many tourist destinations operate marketing and/or management offices that seek to maximize economic opportunities from tourism, while minimizing adverse impacts. A recent study of destination marketing organizations in Canada found that, while awareness of the need to reduce greenhouse gas emissions amongst Chief Executive Officers is relatively high, more progress needs to be made in addressing climate change (Dodds, 2010).

Mitigation efforts in some destinations have progressed strongly on the back of nature-based tourism, and thrived off an image of environmental sustainability. One such example is New Zealand. For the past ten years Tourism New Zealand has marketed the country using the award-winning (50 awards in total) 100 per cent Pure New Zealand campaign (Plate 6.1). However, the success demanded action to underpin the integrity of the brand, especially in the light of growing criticism related to carbon dioxide emissions from aviation. As Tourism New Zealand – the owner of the brand – put it pointedly, '100 per cent Pure sells the sizzle — but where's the sausage?' In response to these concerns, Tourism New Zealand and its partners engaged in the climate change issue by developing a Climate Change Action Plan, by implementing the Qualmark Responsible Tourism certification scheme and by improving the performance of their own corporate activities. Monitoring visitor satisfaction on environmental issues, including climate change initiatives, is an important part of Tourism New Zealand's market research. The New Zealand Tourism Industry Association, in partnership with the New Zealand

PLATE 6.1 Mountain landscape around Queenstown and Glenorchy, often used in Tourism New Zealand imaging and branding. (S. Becken)

Energy Efficiency and Conservation Authority, is currently in its third stage of undertaking energy audits of tourism businesses and disseminating energy efficiency information to the sector. Most recently, many of the above-mentioned initiatives have experienced a decline in funding and government attention as a result of a new government having different priorities. This experience highlights that government programmes related to climate change mitigation are often at the whim of political trends and opinions.

Case study of leadership in a community

As is also the case in New Zealand, some Canadian tourist destinations believe there is an opportunity to leverage themselves as an environmentally friendly destination by implementing climate change measures (Dodds, 2010). In other instances climate change initiatives implemented by local communities for reasons other than tourism development have boosted tourist arrivals. One such example is the city of Växjö in Sweden. Here, change has been driven by the local people and leaders in local businesses, benefiting government and tourism as a result.

Växjö, located in the south of Sweden, has a population of about 82,000 people. The area comprises large forests and 200 lakes. In 1996, encouraged by the local population, the city decided to become a fossil-fuel-free community by 2030. By 2009 CO_2 emissions had decreased by 34 per cent compared with 1996 and more than half of the city's energy demand was met by renewable energy sources (renewable electricity, wood and other biomass). At the same time, economic growth increased by 70 per cent. Växjö has received many awards and now brands itself as 'the Greenest City in Europe'.

As a result of this leadership, and the demonstration effect of the many projects that have been implemented, the city now attracts a large number of people and media who wish to learn about the different sustainability initiatives. The citation for the World Clean Energy Award, which was given to Växjö in 2007, explicitly states that not only have the immediate population benefited from the initiatives, but so have visitors to the city and the wider area.[1] The main innovations that underpin the success include:

- Making the most of the natural resources that the area is endowed with. In this case, the city is benefiting from its large forestry industry and has become a pioneer in using forestry waste for energy production. Two thirds of Växjö's heating comes from this source.
- The use of innovative investments and financial tools. The ecoBUDGET was introduced to allow inclusion of environmental considerations and greenhouse gas reduction initiatives in budgeting and planning processes.
- Reducing energy use through energy efficient street lighting, the installation of solar panels, the promotion of locally grown food, the construction of cycle paths and the provision of rental bicycles.

As shown in Figure 6.1, all of these initiatives benefit tourism while reducing emissions. While they were driven by the city leaders and the local population, a key ingredient for success was the strong financial and technical support from a wide range of sources. The European Commission, for example, provided financial support for the installation of photovoltaic panels, ICLEI Local Governments for Sustainability supported the project as a case study and funding was received from the European ANSWERS project and the 'Interreg IVB North Sea Region' (a project funded by the European Union).

(a) (b)

FIGURE 6.1 (a and b) Cycle paths and tourist attractions in Växjö, Sweden. (www.vaxjo.se/upload/33088/vaxjo_runt_eng.pdf)

Case study: Carbon-neutral destinations

In many places initiatives to reduce GHG emissions are driven by the tourism sector, rather than the community. This contrasts to the above example from Sweden. A key motivation for such initiatives relates to a (potential) competitive advantage in relation to increasingly environmentally aware tourists (see also Section 6.5). Hence, an increasing number of destinations have been announcing plans to become 'the world's first carbon neutral destination' (see also Gössling, 2009). Examples include Costa Rica, Bonaire, Sri Lanka, the Seychelles and Egypt's mass tourism destination Sharm El-Sheikh. Despite great ambitions, none of the destinations have achieved carbon neutrality to date, and in most cases detailed information on how they are going to reach this status is not available. Two destinations, however, appear to have made some progress: the island nation of the Seychelles and Kangaroo Valley in Australia.

The Seychelles has recently launched the world's first 'carbon neutral natural reserve', Cousin Island. Following a carbon footprint assessment, the island found that it was responsible for the emission of more than 1,500 tonnes of CO_2 equivalents annually. This included travel to the island by tourists. A forest restoration project on the island was estimated to offset a proportion of these emissions, but additional carbon credits had to be purchased from overseas to make up the difference, in this case a verified clean cook stove project in Darfur, Sudan. This initiative is part of a wider strategy by the Seychelles (for further details see Gössling and Schumacher, 2010).

Kangaroo Valley lies between Sydney and Canberra, Australia. It is a popular tourist destination, being described by many as 'the most beautiful valley in Australia'. The local Kangaroo Valley Tourist Association has conservation of the Valley at the core of its mission, which includes concerted actions to reduce GHG emissions. A code of ethics has been developed. It implicitly acknowledges the United Nations Local Agenda 21, the UNWTO's Global Code of Ethics for Tourism, the Cape Town Declaration for Responsible Tourism and the Davos Declaration on Climate Change and Tourism. In 2010 Kangaroo Valley gained international recognition by receiving Highly Commended in the 'Best Destination' category in the World Responsible Tourism Awards.

When the initiative started in June 2007 about one third of the tourism operators joined. Since then membership has escalated. The Association estimates that the overall carbon footprint reduced by 20 per cent between 2008 and 2010. Most of the members have audited their carbon footprint using the Association's carbon footprint calculator. The Kangaroo Valley Tourist Association is using a Green Kangaroo symbol to indicate the tourist operators who are participating in the 'Reducing our Footprint' programme. Initiatives by individual businesses are also highlighted on the Association's web page. One example is the Kangaroo Valley Horse Trail. The company is setting up its own hydroponics growing area. This will be run on solar energy so that their 38 horses can be fed without having to freight in feed. This will also reduce the economic impact of higher feed prices during times of drought. Another example is Crystal Creek Meadows, a business that promotes low carbon activities such as biking and walking (Plate 6.2). Crystal Creek Meadows has to date won two national tourism awards, the Qantas Award for Excellence in Sustainable Tourism 2009 and the Climate Action Award 2010. They have reduced their carbon footprint from 50 tonnes to 14 tonnes of CO_2 and have also embarked on a series of climate change adaptation measures.

The critical success factor for this initiative is that participation is voluntary, and all businesses are self-motivated. Future initiatives, such as the introduction of so-called producers tours to increase the use of local produce in menus and for self-catering holidays, will further increase community cohesion and stimulate local

PLATE 6.2 Biking in Kangaroo Valley, Australia. (Man of Kangaroo Valley)

economic development. More recently the Association changed its message from 'becoming carbon neutral' to 'reducing our footprint'. This is in recognition of the huge challenge associated with wider supply-chain emissions and the transportation of tourists, and to avoid overstatements and a risk of 'green-washing'. Importantly, the Association is now considering addressing GHG emissions from transport to the region.

Other examples of destinations reducing emissions

Many destinations are actively involved in climate change mitigation, mostly through the use of renewable energy, some of which are listed below. A relevant example is the city of Masdar in Abu Dhabi. This US$22 billion initiative aims to create the first zero-carbon and zero-waste municipality, housing more than 50,000 people. The Abu Dhabi Future Energy Company (which services Masdar City) is also involved in a number of tourism projects. At the 2010 ITB (Internationale Tourismus Börse), the world's largest travel show held annually in Berlin, two representatives from Masdar City shared their experiences and reported progress to date. While little tourism-specific information on Masdar City appears to be publicly available, the more general website reports that the city will be powered by more than 200 MW of on-site renewable energy systems. The use of solar, geothermal and hydrogen energy will power, heat and cool the city, while a desalination plant will provide fresh water. Masdar City will be car-free and feature – at least on paper – a clean, efficient transport system, including a so-called Personal Rapid Transit system with electric 'pods'. The planned innovative transport systems will be of great interest to many tourism destinations.

Destination	*Initiative*	*More information*
Bahias de Huatulco, Mexico	The destination comprises 36 beaches and a wide range of resorts. A Green Team has been established to lead an Agenda 21 Committee. Energy consumption is 64% better than destination best practice, and travel and tourism accreditation is 74% better than best practice.	www.earthcheck.org
Bonaire, Caribbean	Bonaire tourism is measuring and reducing its GHG emissions where possible and offsetting the remaining emissions through renewable energy, energy efficiency or reforestation projects. Bonaire plans to become a leading carbon neutral destination in the world.	www.tourismbonaire.com/en/over-bonaire/sustainable-bonaire
Lukla, Nepal	Lukla is a town in Nepal whose main source of income is tourism. A 100 kW micro hydro plant now provides electricity to 193 households of which 147 have been converted into hotels and lodges.	www.undp.org.np/uploads/newsletter/201005060814551.pdf
Destination Melbourne, Australia	This tourism marketing organization has developed many creative ways to reduce its carbon footprint. One initiative is the inclusion of public transport and bicycles into the salary package of staff.	www.earthcheck.org
Kaikoura, New Zealand	Kaikoura was the second community in the world to reach Green Globe 21 destination benchmark status. Community for Climate Protection, Energy Efficiency, and Trees for Travellers are three relevant initiatives.	www.kaikoura.co.nz

Destination	*Initiative*	*More information*
Queenstown, New Zealand	Queenstown Lakes Biodiesel Consortium provides access to a supply of B20 blended biodiesel for commercial vehicles, mainly from the tourism sector. The volume of biodiesel consumed by operators has doubled each month since the start of a pilot programme in March 2010.	www.eeca.govt.nz/sites/all/files/implementing-commercial-scale-biofuel-in-queenstown-04-10.pdf
Koh Chang, Thailand	'Climate-friendly' island initiative, as part of the Climate Protection in Tourism project, in partnership with Deutsche Gesellschaft für Internationale Zusammenarbeit (GIZ) and Thailand's Designated Areas for Sustainable Tourism Administration (DASTA). The objective is to integrate climate change into tourism planning and management.	www.tatnews.org/Green-Tourism/5309.asp
Oslo, Norway	City bikes for tourists – for rent on a 24-hour basis. Bikes can be picked up from any of over 100 bike stations in Oslo (Plate 6.3).	www.visitoslo.com/en/the-city-bikes-are-here.49108.283193r4d6.tln.html

PLATE 6.3 City bikes for tourists in Oslo, Norway. (S. Becken)

6.3 Transportation

Background

Transport is a key driver of energy use, making up about one quarter of global demand (OECD, 2001). Forecasts indicate that transport-related CO_2 emissions will increase by up to 80 per cent by 2030, compared to 2002 (IPCC, 2007b). The CO_2 emissions associated with tourism-related transport alone are currently in the order of 981 Mt (UNWTO *et al.*, 2008). More than half (52 per cent) of these are estimated to be associated with air travel, while 43 per cent of emissions are from cars, and 5 per cent from other transport modes. International tourism makes up 44 per cent of the total 981 Mt CO_2 emissions.

Reducing emissions from transport has proven to be extremely challenging. Decarbonizing transport and replacing the current technology of combustion engines is much more costly than decarbonizing electricity production (Peeters, 2010). This makes transport-related mitigation measures economically and politically uncompetitive. The share of emissions that can be abated at a cost of less than US$50 per tonne of CO_2 is about 30 per cent for road transport and 12 per cent for air transport, compared with 75 per cent for electricity (Peeters, 2010).

Another reason for the difficulties in reducing transport emissions is that mobility and car ownership are commonly associated with substantial psychological and social benefits (Moriarty and Honnery, 2008a, 2008b). Views of tourism 'as a basic right' highlight the value people attach to travel (Becken, 2007; DEFRA, 2007). Due to the importance of personal mobility in modern society, most mitigation measures are focused on new technologies or traffic management, rather than on the reduction of demand (Table 6.2). More information on different transport technologies, fuels and management options is provided in Becken and Hay (2007).

A number of studies have demonstrated – through scenario analyses – that technological changes alone will be insufficient to reduce emissions to required levels (e.g. to keep anthropogenic warming below 2°C), and more fundamental

TABLE 6.2 Overview of mitigation technologies and practices from the transport sector (non-technological measures such as lifestyle changes are not included). (After IPCC, 2007b: 10)

Mitigation technologies/practices currently available	*Mitigation technologies/practices available by 2030*
More fuel efficient vehicles, hybrid vehicles, cleaner diesel vehicles, biofuels. Modal shifts from road transport to rail and public transport systems, non-motorized transport (cycling, walking). Land use and transport planning.	Second or third generation biofuels. Higher efficiency aircraft (see below). Advanced electric and hybrid vehicles with more powerful and reliable batteries.

transformations are required if transport is to make a real contribution to reducing the rate of global climate change (Bristow *et al.*, 2008; Kwon, 2005; Moriarty and Honnery, 2008a). This is particularly so for aviation, where growth rates are likely to outweigh efficiency improvements (Dubois *et al.*, 2011; see also Chapter 5). Most of the studies predicting growth in passenger demand assume increasing economic prosperity. Growth in incomes is the key driver for people making more frequent and longer trips (Mayor and Tol, 2010). Macintosh and Wallace (2009) acknowledge that major recessions, increasing oil prices, or both, could invalidate such scenarios (Becken, 2008a, 2011a). The remainder of this section provides a detailed discussion of mitigation measures and potentials in the aviation sector, followed by case studies related to other transport forms, namely train travel, cycling and cruises.

Reducing emissions from aviation

Aviation is the 'Achilles heel' of tourism when it comes to climate change (see also Chapter 5 for policy dimensions). The large climate impact of air travel results from both carbon intensity and long distances travelled by plane. In Europe, for example, air travel is the main transport mode for about 20 per cent of trips, but it accounts for 55 per cent of passenger kilometres and 50 per cent of tourism CO_2 emissions (Bows *et al.*, 2009). The aviation sector has made great improvements over the last decades: engine fuel consumption has improved by 49 per cent on average between 1960 and 2010 (i.e. about 1 per cent per year, Sgouridis *et al.*, 2010), while aircraft fuel burn per seat decreased by 82 per cent (IATA, 2009a). But how much does aviation really contribute to climate change? There are a number of ways of looking at this, with different sources providing slightly different findings (Table 6.3). Generally it is estimated that aviation contributes between 2 and 3.5 per cent of anthropogenic global warming. Not all aviation is related to tourism.

Emissions from aviation are forecast to grow due to increased passenger demand, by 111–144 per cent between 2005 and 2025 (Macintosh and Wallace, 2009). However, a range of measures are expected to slow this 'business as usual' rate of increase. These include energy efficiency improvements of the aircraft themselves, air traffic management and low carbon fuels (e.g. biofuel). In the short term,

TABLE 6.3 Aviation's contribution to anthropogenic climate change.

Measure	*Data (%)*	*Source*
Aviation's share of global CO_2 emissions	2	IATA (2009a)
Aviation's consumption of global fossil fuels	2–3	Lee (2010)
Relative contribution of aviation to global anthropogenic carbon emissions	3	Sgouridis *et al.* (2010)
Total radiative forcing (including gases other than CO_2) from aviation	3.5	Lee *et al.* (2009)

emission reductions from technological (aircraft-related) developments dominate the reduction path of CO_2 emissions from aviation. But in the longer term there are great hopes for reductions as a result of low carbon fuels. Not surprisingly, reductions in demand do not feature in the IATA's management of future emissions. Importantly, several studies have found that only 'aggressive measures' would stifle demand. For example, a price of US$200 per tonne of CO_2 would reduce emissions by 8 per cent relative to a business-as-usual scenario (Sgouridis *et al.*, 2010). Non-demand measures are discussed in more detail below.

> In order to survive high fuel costs, the aviation industry has to fill aeroplanes to be as energy- and cost-efficient as possible. Consolidation, route rationalisation, more efficient operational procedures and low-cost competition will tend to moderate prices to enable airlines to fill seats, especially on popular routes. (Francesca Ecsery, General Manager, Cheapflights.co.uk)

Technological improvements

Airlines have a strong interest in reducing GHG emissions. Fuel use contributes up to 40 per cent of operational costs. In the face of increasing global fuel prices any savings in energy consumption help the economic bottom line, especially given the extremely competitive environment in which airlines operate. Aircraft size is one important factor in improving emissions per available seat kilometre. A recent study found that small gains (less than 1 per cent per annum) can be made by employing larger aircraft, especially on short- and medium-haul routes that currently employ smaller single-aisle aircraft (Morrell, 2009).

Many other options are available to improve the airframe and the engine. Technical details are widely available, for example through special reports by the IATA or by consultants (e.g. Peeters *et al.*, 2005) (Table 6.4). In its Technology Roadmap Report, the IATA (2009b) provides information about saving potentials with current technologies and future developments. Most individual measures, such as graphic films applied to the exterior of aircraft, or engine retrofits, save about 1 per cent of fuel consumption. However, raked or blended wingtips have the potential to save 3–6 per cent of fuel.

Operational measures

Already in 1999, the IPCC estimated that there was a 12 per cent inefficiency in air traffic management strategies such as continuous descent approaches (Penner *et al.*, 1999). The IATA (2009a) reports that since this seminal IPCC report, 4 per cent efficiencies have been achieved. Thus there is still considerable potential

TABLE 6.4 Emission reductions with technologies currently available. (After IATA, 2009b)

Airframe improvements		*Engine developments*
• Active load alleviation • Aircraft graphic films • Advanced alloys • Blended winglet • Composite primary and secondary structures • Drag reduction coatings • Fluoropolymers • Friction stir welding • Glar • High strength glass microsphere • Zonal dryer	• High power light-emitting diode (LED) for cabin lighting • Landing gear drive • Laser beam welding • Lithium batteries for secondary power • More efficient gas turbine auxiliary power unit • Raked wingtip • Variable camber with existing control surfaces • Wingtip fence	• Advanced combustor • Engine retrofits: ◦ Advanced heat-resistant materials ◦ Better blade design ◦ More efficient energy management • Variable geometry chevron • Variable fan nozzle

to reduce emissions. For example, the current fragmentation of the European airspace results in an average of 50 km unnecessary flying. The Single European Sky (SES) is a policy to address this and streamline air traffic management to reduce emissions, delivering environmental as well as economic benefits. The updated policy framework contains a 'network approach through performance targets … a reorganization of air space in cross-border air space blocks and the flexible use of air space in coordination through civil and military use' (European Community, 2009). Other examples of improved air traffic management include the Next Generation Air Traffic Management (ATM) system in the United States (the potential of a 57 per cent reduction in delays), the Pearl River Delta ATM system in Hong Kong and RVSM (Reduced Vertical Separation Minima) over Russia (IATA, 2009a). The investment required to enable these improvements amounts to US$58 billion. More detailed technical information is available in IATA (2009b).

A current initiative is the San Francisco Oceanic Tailored Arrivals (OTA) Trial. This is a joint venture between Boeing, the National Aeronautical and Space Administration (NASA), the Federal Aviation Administration (FAA) and San Francisco Airport (United States). The initiative allows aircraft to make full use of interlinked on-board and ground technology to descend into the airport with minimal direct Air Traffic Control intervention. An aircraft flying an OTA procedure begins a very slow gradual descent while still hundreds of miles from the airport instead of flying a series of level segments. Air New Zealand, the first non-American airline to undertake OTA, reported to have saved more than 24 tons of jet fuel and reduced CO_2 emissions by more than 76 tonnes through OTAs. During a 12 month trial the airline was able to use the full tailored arrival procedure around 30 times and a partial procedure about 100 times.

Another interesting project is the Asia and South Pacific Initiative to Reduce Emissions (ASPIRE, 2010). This brings together Airservices Australia, Airways New

Zealand and the FAA. ASPIRE is designed to raise the industry's environmental performance by reducing fuel burn and emissions by exploring improvements to operational procedures in all aspects of flight to improve on current best practices and existing technologies. A test flight by Singapore Airlines in February 2010, from Los Angeles to Singapore via Tokyo, operated under optimum air traffic conditions and achieved savings of 6 per cent compared with a normal flight.

It is important to note that, while air traffic management plays an important role in climate change mitigation by aviation, the overall growing passenger volumes will increasingly challenge such initiatives. EUROCONTROL (2008) estimates that demand for flights within Europe will exceed airport capacity by 0.9 million flights in 2025, rising to 2.3 million un-accommodated flights by 2025. On the other hand, such pressure on capacity will provide additional incentives to optimize operational systems.

Alternative fuels

Airlines are increasingly seeking alternative fuel sources such as 'sustainable biofuels'. The IATA (2009a, 2009b) believes that second- or third-generation biofuels (e.g. algae, Jatropha, camelina) could reduce CO_2 emissions from aviation by 80 per cent on a full carbon life cycle basis. Many airlines are involved in biofuel research initiatives and have set themselves targets for its use. For example, Virgin Atlantic anticipates that by 2015 up to 5 per cent of its fuel uplift will come from biofuels, and at least 10 per cent by 2020. Similarly, the Dutch carrier KLM (the Netherlands) has announced the formation of a new company to develop sustainable biofuels called SkyEnergy, in a joint venture with North Sea Petroleum and Spring Associates. The WWF will advise this consortium in relation to ecological aspects. In early 2011 British Airways and the US-based Solena Group announced an agreement to build Europe's first plant to produce jet fuel from organic waste. Subject to regulatory approval, the plant is expected to go online in 2014, and produce about 72,700 litres (16 million gallons) of fuel annually, all of which will be sold to British Airways. Air New Zealand is involved in research on biofuel production from algae and plans to use one million barrels of sustainable fuel annually by 2013.

A series of test flights (Table 6.5) have demonstrated that biofuel can be used to partially replace fossil aviation fuel. However, while biofuel seems promising from a technological viewpoint there remain issues around the sustainability of its production (e.g. taking land from other forms of use, such as food production). Questions have also been asked about the large amounts of biofuel needed to power the worldwide fleet. Peeters (2010) estimated that if the biofuel came from Jatropha, an area the size of Germany, France, the Netherlands and Belgium combined would be required to fuel current air transport (this research refers to all jet fuel in 2005).

It is clear that many airlines are actively involved in reducing fuel consumption and GHG emissions. Most of their websites provide detailed information on environmental initiatives, corporate responsibility and links to research projects and

TABLE 6.5 Biofuel flights undertaken by various airlines. (After Steele, 2009)

Airline	*Aircraft*	*Partners*	*Date*	*Biofuel*
Virgin Atlantic	B7474-400	Boeing, GE Aviation	23 Feb. 08	Coconut and Babassu (20% blend)
Air New Zealand	B7474-400	Boeing, Rolls Royce	30 Dec. 08	Jatropha (50% blend)
Continental Airlines	B7474-800	GE aviation, CFM, Honeywell UOP	7 Jan. 09	Algae with Jatropha (50% blend)
JAL	B7474-300	Pratt and Whitney, Honeywell UOP	30 Jan. 09	Camelina, Jatropha and Algae blend (50%)

carbon offsetting programmes. One good example is Air Mauritius. The company has recently signed an agreement with the Mauritian Wildlife Foundation in support of conserving the endemic flora and fauna of Mauritius. The main initiative in the programme is the 'one takeoff /one tree' campaign, which involves planting one endemic tree for each takeoff of an Air Mauritius aircraft from Mauritius. Under this project, 6,000 endemic trees will be planted each year in Mauritius and Rodrigues to help restore natural habitats and create local awareness.[2] The compilation below provides a snapshot of many other initiatives that have been implemented by airlines around the world. An interactive map on similar initiatives worldwide has been compiled by the Air Transport Action Group.[3]

Company	*Initiative*	*More information*
Thomas Cook, Great Britain	In September 2010 Thomas Cook Airlines became the first UK airline to achieve ISO 14001 certification. The company's website provides a lot of detail on mitigation measures, including its involvement in the Sustainable Aviation Initiative and the maintenance of an Operational Efficiency Working Group.	www.thomascook.com/sustainable-tourism/looking-after-the-environment/
Virgin, Great Britain	Reducing weight: loading on-board catering more effectively, using different products and materials, minimizing waste, encouraging passengers to take less luggage, minimizing overfuelling. Virgin Atlantic is one of the founding members of the Aviation Global Deal Group. It aims to help shape international efforts to develop a sustainable global policy framework for reducing CO_2 emissions from aviation.	www.virgin.com

Company	*Initiative*	*More information*
Lufthansa, Germany	Lufthansa is investing in a range of mitigation initiatives, including a geothermal facility with 63 earth probes at the Seeheim Training & Conference Center. In 2009, the largest aircraft maintenance company in Beijing, AMECO (a joint venture of Lufthansa Technik and Air China), commissioned an environmentally friendly gas-powered heating plant. This will likely increase thermal efficiency by 10% and decrease CO_2 emissions by 50%.	http://verantwortung.lufthansa.com/fileadmin/downloads/en/LH-sustainability-report-2010.pdf
Air New Zealand, New Zealand	Reducing weight: the introduction of lighter weight equipment such as cargo containers, carts and inflight products. The amount of drinkable water carried on each flight has been significantly reduced to align with actual demand. Installing zonal dryers to remove water stemming from passengers' respiration and adding substantial weight from insulation blankets in the crown (ceiling) of the B737, A320, B777 and B767 aircraft, saving 5,400 tonnes of CO_2 per year. The Air New Zealand Environment Trust provides customers with the option to donate to credited carbon sequestration projects (e.g. native reforestation projects).	www.airnewzealand.co.nz/environment
KLM, the Netherlands	Supporting research to develop a new aircraft that is 50% more energy efficient. KLM's contribution includes practical insight and data about aircraft deployment and performance. The aircraft will be ready to fly in 2025. All of KLM's Boeing 737 Next Generation aircraft are equipped with winglets. This has improved aerodynamics and reduced fuel consumption, with CO_2 emissions reduced by about 3% for this aircraft type.	www.klm.com/csr/en/climate/footprint/measures/index.html
Japan Airlines, Japan	By regularly washing off dust and dirt that has attached to the compressor in the engines, compression efficiency can be improved and fuel consumption reduced by about 1% during cruising.	http://press.jal.co.jp/en/release/200910/001341.html

Case study: mitigation by a land transport company

InterCity Coachlines in New Zealand has been involved in climate change mitigation since 1998. In 2007 it announced that its aim was to operate the world's first carbon neutral national transport system. The InterCity Group includes InterCity Coachlines, InterCity Ferries, Newmans Coach Lines, GreatSights New Zealand, Fullers GreatSights Bay of Islands and awesomenz.com. Together these form a significant part of New Zealand's national public transport network. It consists of 170 coach services and three ferry services a day to 600 communities, towns and cities across New Zealand.

The company has initiated a large range of initiatives to reduce GHG emissions, including driver training and the purchase of larger buses that can transport passengers more efficiently on core routes (e.g. Auckland to Wellington) (Plate 6.4). Other major achievements include:

2002 – committing to new vehicles voluntarily meeting latest European emission standards, representing an investment in excess of NZ$20 million.

2003 – applying an anti-fouling coating to the hulls of vessels to improve fuel efficiency. The result was a 10 per cent reduction in fuel consumption, with corresponding reductions in emissions.

PLATE 6.4 One of four double-decker coaches which commenced operations in 2007. It meets Euro 3 Emission Standards. The bus is the largest in New Zealand, and is estimated to carry 23,725 passengers per year – the equivalent of removing 47,000 private vehicle journeys from the roads each year. (InterCity Coachlines)

2004 – removing interior walls at head office to allow more efficient air conditioning and natural light distribution, reducing annual power consumption by 7 per cent.

2006 – upgrading server systems and implemented new server sharing software, a project worth NZ$3.6 million. This reduced server numbers from 13 to three, reducing power requirements by 70 per cent.

2007 – preparing and implementing an emission reduction plan. This resulted in annual energy savings of 15 per cent. It also included a goal for all company cars to reduce emissions by 20 per cent per vehicle within three years through such initiatives as downsizing vehicles.

2008 – moving into a new office, which included an investment of NZ$100,000 for new air conditioning and improved electrical systems to minimize power consumption.

2008 – trialling new fuel filtering technology on vessels aimed at reducing fuel consumption by 12 per cent.

2008 – initiating code sharing between brands in winter when bus and vessel loadings are lighter, resulting in 50 per cent emission reductions for the winter months.

2010 – Fullers GreatSights Bay of Islands received a Qualmark Enviro-Gold Award in recognition of its high standards in environmental practices.

2010 – InterCity Coachlines, Newmans Coachlines and GreatSights New Zealand received a Qualmark Enviro-Bronze Award in recognition of their high standards in environmental practices.[4]

Clearly the InterCity Group has been a leader in implementing emission reduction measures, to the point where they were sometimes ahead of legislation and standards. Regrettably their investment in carbon offsetting has since been terminated, due to the government's introduction of an Emissions Trading Scheme; otherwise this would have resulted in double accounting of emission reductions. There was minimal use of the option provided to customers to offset their carbon emissions, so this is no longer an option in the booking progress. While the Group retains its core values of environmental protection and climate change mitigation, the global financial crisis has placed significant pressure on business operations, to the extent that some environmental initiatives have become more difficult to implement (InterCity Group CEP, personal communication).

Case study: 'Slow tourism' – cycle tourism

Slow tourism or slow travel has been defined as 'an emerging conceptual framework which offers an alternative to air and car travel, where people travel to destinations more slowly overland, stay longer and travel less' (Dickinson and Lumsdon, 2010). Bicycle tourism is an important component of slow travel. The 'Bett & Bike' initiative by the German cycle club Allgemeiner Deutscher Fahrrad Club (ADFC) is a good example of how creative thinking can enhance the experience of tourists (in this case the niche market of cycling tourists), create business opportunities and reduce environmental impacts, including GHG emissions.

ADFC reports the following statistics (ADFC, 2010):

- The number of ADFC certified bike-friendly Bett & Bike accommodation providers has increased from 216 in 1995 to over 5,007 (March 2010).
- The above providers operate in 3,412 different locations across Germany.

The network thus encompasses all of Germany (Plate 6.5). It is the most successful campaign to date tailored to cycle tourists. The ADFC specifies different criteria for certification, depending on whether the participant is an accommodation provider (often bed and breakfast), a restaurant or a campground. For example, accommodation businesses have to provide lockable facilities for bicycles, a room for drying clothes, healthy menus tailored to the requirements of customers, information on local routes and on public transport, repair kits and facilities and information on local bicycle shops. Some operators have developed a local bike map in partnership with the ADFC, and have released joint flyers and press releases to communicate and market their products. Information on cycle

PLATE 6.5 Cycle tourism on small country roads in South Germany. (J. Roberts)

routes is also available online and can be downloaded to GPS and other handheld devices.

Importantly, initiatives such as Bett & Bike encourage cycle tourism and thus contribute to reducing GHG emissions from tourism transport. Further research into the carbon intensity of cycle tourism would be informative, including determining how many tourists use a combination of car and bicycle during a vacation.

Case study: Water-based transport – cruise ship tourism

Water-based tourism has not been studied extensively from a GHG emissions perspective, but there is increasing evidence that it is comparatively carbon-intensive. A recent study on cruise ship tourism in New Zealand (Howitt *et al.*, 2010) estimated a carbon intensity of 390 g CO_2 per passenger kilometre. This is substantially more than economy class travel on an aircraft (for example, Emirates claims that the A380 emits less than 75 g CO_2 per passenger kilometre). Similarly, the study stated that the hotel function of a cruise vessel uses about 1600 MJ per visitor night, 12 times larger than the equivalent value for a land-based hotel. This is partly explained by the high level of luxury on board a cruise ship, but also because of inherently inefficient systems, such as electricity production based on diesel generation. Based on figures from the cruise line Carnival, the carbon footprint of a cruise ship passenger is 36 times greater than that of a Eurostar rail passenger and more than three times that of someone travelling on a standard Boeing 747 or a passenger ferry (Archer, 2008). Ocean cruising globally is estimated to produce about 34 Mt of CO_2 per annum, while civil aviation emissions are about 640 Mt (World Economic Forum, 2009).

Ecoventura, a small cruise operator in the Galapagos Islands of Ecuador, has shown leadership in reducing GHG emissions (The International Ecotourism Society, 2009). The company transports about 4,000 passengers per year on three expedition vessels. One of Ecoventura's vessels, the M/Y *Eric*, was recently modified to reduce the consumption of fossil fuels. It was equipped with 40 solar panels and two wind generators and is the first hybrid tour yacht in the Galapagos Islands (Plate 6.6). The goal is to produce 9 kw per day, or the equivalent of 17 per cent of the energy used to run the on-board generators. The project was financed through a partnership with Toyota, a supporter of the WWF, the Galapagos Marine Biodiversity Fund and other conservation programmes in Galapagos. The company plans to refit its two other vessels with hybrid wind and solar power by 2011.

Ecoventura is also the first carbon neutral tourism operation in Ecuador. In partnership with the WWF, the company established the Galapagos Marine Biodiversity Fund with the objective to strengthen the ability of local communities to manage their natural resources through environmental education and marine conservation.

PLATE 6.6 Solar panels on an Ecoventura cruise vessel in Ecuador. (www.ecoventura.com/home.aspx)

Other examples of transport related mitigation

The following examples illustrate how different transport providers address the issue of energy use and GHG emissions. They highlight initiatives in the rail transport sector, tour travel and other transport providers, such as scenic flights or rental cars.

Company	*Initiative*	*More information*
Swedish Railways, Sweden	Provides an emission calculator, as well as environmental reports of emission avoidance to corporate customers (i.e. companies that are frequent customers); separates waste on board the train and uses compostable packaging.	www.sj.se
Dutch Railways, the Netherlands	Optimized brake energy recuperation settings and saved 1% of total electricity use (pay-back time of four weeks).	Peeters (2010)

Company	*Initiative*	*More information*
SNCF Railways, France	Promoting 'ecomobility' including innovative concepts for door-to-door travel (optimizing travel chain) and hire of electric taxis.	www.sncf.com/en_EN/html/page/CH0006-Ecomobility/BR0535-Travelling-green.html
Deutsche Bahn, Germany	'Fahrtziel Nature' – in partnership with major nature conservation organizations, the DB offers attractive train packages to 19 unique nature destinations (e.g. German's Nationalparks) (Plate 6.7).	www.fahrtziel-natur.de
EuroVelo, Europe	EuroVelo, the European cycle route network, is a project of the European Cyclists' Federation (ECF). EuroVelo is made up of 12 routes, totalling over 66,000 km, of which about 45,000 km are already in place.	www.ecf.com/14_1
A Safe Passage, Costa Rica	Bus and shuttle service in Costa Rica run on biodiesel; also for every passenger pickup a tree is planted to further offset emissions.	www.costaricabustickets.com/
Electric buses in Hanoi, Vietnam	A fleet of small electric battery-powered buses, seating up to eight passengers, provide tours of the old quarters in Hanoi.	www.saigonblues.com/saigon-travel/electric-bus-for-ha-noi-city-tours-haivenu-tours/
Green Island and Siaoliouciou, Thailand	The Thailand Tourism Authority is investing approximately NT$800 million (US$25 million) in the next three years to replace about 14,600 regular scooters with electric scooters by 2013.	www.taipeitimes.com/News/taiwan/archives/2010/05/17/2003473138
Hassle-free Tours, New Zealand	Using biodiesel blend (B20) for their fleet of trucks and jeeps and achieving a 15–18% reduction of GHG emissions. Planning to move to pure biodiesel in summer months as the temperature is then more favourable for using such a fuel.	www.eeca.govt.nz/node/9214
Real Journeys, Te Anau, New Zealand	Fuel consumption of aircraft was reduced by 2 per cent as a result of the manual recording of fuel consumption on scenic flights and posting the information on a noticeboard.	www.eeca.govt.nz/sites/all/files/flight-operators-fuel-efficiency-article-06-10.pdf

Company	*Initiative*	*More information*
Green Tomato Cars, UK	Uses Toyota Prius as taxi services. They also suggest that passengers share a car where possible, offset emissions, and assess drivers for low carbon driving techniques.	www.greentomatocars.com/
Big Cat Green Island Cruises, Australia	Great Barrier Reef operator (more than 10,000 passengers p.a.) with a zero carbon footprint. For example, a catamaran was cut in half and extended by five metres to add 100 more seats.	Zeppel (2011)
La Petite Reine, France	Specialist in the intra-urban final delivery of goods (e.g. food) using a pedal powered 'cargocycle' with a capacity of up to 180 kg.	www.lapetitereine.com/uk/index.php?PHPSESSID=04e19f277d69b063f7fc4c211116cd2d

PLATE 6.7 Fahrtziel Natur advertisement by the Deutsche Bahn. (S. Becken)

6.4 Mitigation at facilities and in operations

Background

The tourism facilities and operations sector is large. It comprises tourist accommodation, attractions, activity operators, travel agents, tour operators, wholesalers and other businesses involved in providing services to tourists or intermediaries. As operators in this sector are also part of the wider business community (i.e. the commercial sector), most providers are likely to have encountered information on climate change mitigation in one way or another. There is a wide range of initiatives that address business sustainability, corporate governance, environmental management, best practices, responsible resource use and so forth. Many of these are not tourism specific, but there are also many international or national programmes directed explicitly at tourism businesses.

The accommodation sector, most prominently the hotel industry, has been highly proactive and focused in developing initiatives to reduce energy consumption as well as GHG emissions. While concern about the global climate might be a motivating factor, tourism businesses are also influenced by the financial and other risks inherent in volatile energy prices, blackouts, energy policy changes, carbon taxes and negative consumer perceptions. Typically energy would represent around 5 per cent of the operating costs of a hotel, second only to labour costs (e.g. Bohdanowicz *et al.*, 2001). The EnergyStar programme of the Environmental Protection Agency in the United States emphasizes the need to communicate with hoteliers in a language that matches the other financial metrics they use. For example, energy reductions improve operating income, which translates to higher asset values for the owner. Also, energy consumption can be juxtaposed to revenue per available room (RevPAR), and savings can be interpreted as increased revenue per room. A 10 per cent energy saving for a typical American full-service hotel would be the equivalent of an increase in the return on the daily room rate of US$1.35 (EnergyStar, no date).

Typically the largest energy end use is heating and air conditioning (e.g. Bohdanowicz and Martinac, 2007; Trung and Kumar, 2004), although uses such as the supply of hot water, refrigeration and lighting are also significant contributors (Keesman *et al.*, 2006). A life cycle analysis of hotels in the Balearic Islands found that the operating phase of a hotel accounts for 70–80 per cent of the energy used over the lifetime of the hotel (Rossello-Batle *et al.*, 2010). Hence, when attempting to measure the total energy requirements of the accommodation industry, it is important to also include the construction and demolition phases. Every accommodation business is different – accordingly it is very difficult to compare the energy use of different providers. Accommodation type (e.g. hotel versus hostel or campground, Becken *et al.*, 2001) plays an important role, but so do factors such as the design, age and size of the building(s), time since the last major retrofit, climatic conditions, type of energy systems, operational and maintenance schemes, energy regulations and costs (Priyadarsini *et al.*, 2009), occupancy levels, customer preferences, culture and awareness of resource use (Bohdanowicz and Martinac, 2007).

A number of studies have investigated ways to reduce energy use in tourist accommodation. New technologies explored for tourism include, for example, heat pumps for hotel swimming pools (Chan and Lam, 2003), deep ocean cooling for air conditioning systems (WWF *et al.*, 2010) and solar control window film to reduce the need for air conditioning (Chan *et al.*, 2008). There are abundant case studies and examples of energy savings measures in the accommodation sector, of which a few will be highlighted further below. Overall, it appears that the savings potential could be between 20 and 40 per cent compared with current usage (Becken, in press; Duffy-Mayers, 2010).

Researchers have also explored the potential for renewable energy sources for tourist accommodation (e.g. Bode *et al.*, 2003 Plate 6.8). A recent study by Dalton *et al.* (2008) researched the feasibility of renewable energy supply (RES) for a large-scale resort in Queensland, Australia. A simulation showed that a 100 per cent RES was both feasible and economically viable, although a hybrid approach to RES, including a diesel generator, provided the lowest net present-value costs. Taking into account that hotels require 24 hour operation, high levels of comfort provision and high reliability, the results showed that wind energy was more viable than using photovoltaic systems. Renewable energy sources are especially attractive for island destinations, where energy is typically imported at great cost (Uemura *et al.*, 2003; Michalena *et al.*, 2009).

Many other businesses involved in tourism have analysed their energy consumption as well as emission reduction potentials. One pertinent example is the

PLATE 6.8 Solar hot water on tourist accommodation in Suva, Fiji. (S. Becken)

conference and convention industry, and events organizers more generally. The Greener Events Guide (Harding and Malone, 2005), for example, provides a very useful list of opportunities for reducing GHG emissions associated with an event. Items include choice of venue (e.g. good access via public transport, in-house environmental policy and priorities, and rental equipment), travel related emissions (e.g. informing delegates about public transport, promote car sharing, etc.), minimizing air conditioning, choice of food and the content of delegate packs.

Case study: Self-catering tourist accommodation

The Hytte in Northumberland, England (www.thehytte.com), provides self-catering accommodation for eight guests. It is an excellent example of sustainable tourism. Not only has it implemented many of the 'standard' measures to reduce its environmental impact, including the carbon footprint, but it also pioneers 'outside the box' measures. The Hytte is also a very good example of environmental actions being integrated with social responsibility (Plate 6.9).

One of the boldest measures is providing guests with information on their electricity consumption. The owners of the Hytte recorded their energy usage weekly for two years and were therefore able to determine their 'average' energy use. Based on this, an internal electricity meter now allows guests to monitor how much energy they are using during their stay and to then, ideally, attempt to reduce

PLATE 6.9 The Hytte in Northumberland, self-catering accommodation (with turf roof). (www.thehytte.com)

it if they are 'above the average'. Guests who use above 'average' electricity will be charged for excess units in the form of a 'pay back'. This 'pay back' is donated to the Northumberland National Park Environment Association and to the Great North Air Ambulance.

The Hytte also actively encourages its guests to make a contribution by:

- separating recyclable, compostable and cooked kitchen waste in the bins provided, as well as scraps of bread, pasta, rice and potatoes. The latter are fed to the hens on the property, with guests being told their waste is 'recycled into eggs';
- switching off appliances, lights and taps when not in use, and reporting dripping taps;
- supporting local producers and services – 'Eat Local, Buy Local and See Local';
- using the washing lines or an indoor airer, in preference to tumble drying;
- not leaving doors and windows open in winter and closing curtains/blinds at dusk to minimize heat loss;
- having a day on site or using the Roman Wall Bus to reduce car use;
- not throwing away empty complimentary shower gel or cleaning bottles – these are refilled and reused; and
- reducing water and energy consumption by taking a shower instead of a bath.

The Hytte provides comprehensive information on environmental initiatives on its website, including those related to energy savings:

- Refrigerators and freezers are eco-labelled as 'A' rated. Both are switched off when not in use, and the owners ensure that door seals are tight and doors close easily.
- All light bulbs are low energy throughout the Hytte internally and externally, reducing the amount of energy used and minimizing light pollution.
- Zoned thermostats have been installed to control heating in all areas.
- Minimum standards for insulation have been exceeded and the Hytte grass roof provides further insulation.
- All windows are double-glazed with energy efficient 'E' glass.
- Electricity is renewable 'Green Energy'.
- All heating is supplied by a ground source heat pump which provides three units of heating for every one unit required to drive the pump.
- The original thin polystyrene lid on the hot tub was replaced with one posessing far greater insulating qualities. The original polystyrene has been reused by placing it under and around the tub to improve insulation further. A floating cover has been installed to further prevent heat loss.
- The hot water tanks are well insulated and set at 60°C.
- All laundry is washed on site and line dried.

- The Hytte plans to install a 4 kW ground-mounted photovoltaic system in 2011.
- 1,000 native trees and shrubs have been planted to encourage a habitat for wildlife.

Case study: A tourist activity – skiing

Whistler Blackcomb is a major ski resort located in British Columbia, Canada. In terms of size it is by far the largest ski resort in North America, and it features the Peak 2 Peak Gondola (Plate 6.10) which holds the record for the highest and longest unsupported cable car span in the world. About two million skiers visit the resort each year. Given the size and importance of Whistler Blackcomb in international skiing, its environmental leadership is important and likely to be replicated elsewhere.

Whistler Blackcomb was recently awarded the Golden Eagle Award for Overall Environmental Excellence by the National Ski Areas Association (NSAA), and was a finalist in three Silver Eagle categories, namely Excellence in Energy Conservation/Clean Energy, Excellence in Environmental Education and Excellence in Stakeholder Relations. The most recent energy related project is the micro hydro renewable energy project on the Fitzsimmons River. Detailed information on the seven-year process to obtain permits and local support is available on the corporate website. The project is planned to return an estimated 32 GWh

PLATE 6.10 Whistler Blackcomb gondola. (P. Morrison)

of electricity to the grid annually, the equivalent of energizing 3,000 to 4,000 homes, and also equal to Whistler Blackcomb's annual energy consumption.

Whistler Blackcomb developed a specific energy strategy which is part of its wider Sustainability Policy. That was created with input from staff and endorsed by the Senior Leadership Team. A company-wide reduction target of 5 per cent in 2010/11 for electricity and natural gas usage is tied to annual performance reviews and financial incentives. A full-time Energy Manager/Environmental Coordinator supports and oversees all energy reduction initiatives. Examples of mitigation measures include:

- reducing the fleet and vehicle size for higher efficiency;
- purchasing 39 low-emission snowmobiles, representing half of the fleet, and a commitment to purchase additional low-emission replacement models;
- driving General Motors hybrid technology pickup trucks;
- establishing an employee carpooling programme that saves more than 617,000 kg of CO_2 emissions annually;
- retrofits that reduce the annual electricity consumption by 4,575,000 kWh (the power required to run 450 Canadian homes for a year);
- retrofits to buildings that have reduced CO_2 emissions by more than 850 tonnes;
- replacing 11,000 light bulbs with more efficient and environmentally friendly types – the light bulb project saves 1,533,000 kWh per year;
- installing a 1.0 megawatt run-of-river turbine in Flute Creek, to power the lights and provide heat in the lift hut and washrooms at the base of one of the chairlifts.[5]

Case study: GHG emission reductions by a tour operator

Tour operators have an important role in climate change mitigation. The 'Tour Operators Initiative for Sustainable Tourism Development' is an example of an undertaking to improve the environmental performance of tour operators. It is anchored around corporate social responsibility (CSR), and is a joint collaboration of four big and some medium-sized tour operators, supported by UNWTO, UNESCO and UNEP. The aim is to advance coordinated CSR activities in the sector. However, as Zotz (2010) notes, the effectiveness of tour operator initiatives to reduce their emissions is limited, due to fierce competition among operators and a limited preparedness of the customer to support relevant measures. Examples of specific measures by TUI AG, First Choice Holidays, Thomas Cook Group, Rewe Touristik, Kuoni Travel Holding, Hotelplan AG and Alltours Flugreisen are provided in Zotz (2010).

Intrepid Journeys, headquartered in the United Kingdom, makes for another informative case study that illustrates what a tour operator can do to reduce its CO_2 emissions. The issue of climate change and carbon footprint reduction is highlighted prominently on its website.[6] Educational material is provided to customers, including material explaining climate change (with useful links, including to the IPCC website) and carbon offsetting (e.g. certification and standards) along with a video clip. 'Personal tips for reducing emissions' are provided in an interesting and topical manner – for example, 'green Christmas tips' are included on the website at the appropriate time.

The Intrepid website sent a signal that climate change is a significant problem and that the company is taking it seriously. In December 2006, Intrepid Travel announced the goal of becoming a carbon neutral company in 2010. Intrepid underwent a life cycle assessment (LCA), primarily to help identify the main sources of carbon emissions from its operations. Through this process three sources of emissions were identified: corporate emissions (offices and retail stores globally); tours; and the flights that passengers take to join a tour.

Air travel by clients is the largest of the three sources of emissions. For this reason optional carbon offsetting is provided as part of the booking process. In the first year (2007), Intrepid travellers offset 18,700 tonnes of CO_2. The second largest factor is the tours offered. Here, Intrepid claims to consider the following, amongst others:

- Local services – use of locally owned and operated service providers to reduce carbon-intensive supply chains.
- Local transport – use of local public transport wherever possible to reduce fuel usage per passenger.
- Local food and local goods – consumption of locally produced food and goods where possible. This reduces the 'embodied energy' – energy consumed through the production and transport of the food and goods consumed by customers.
- Water bottles – encourage passengers to refill water bottles from water 'bubblers', where available, to avoid unnecessary purchasing of bottled water and subsequent waste disposal issues.

In addition to the water in bottles, the production of a one litre plastic bottle takes two litres of water and 200 ml of oil. Therefore, for every litre of water sold, three litres of water are used. The emissions resulting from the tour (i.e. transport, accommodation and waste) are always offset. The cost is included in the tour price. Information on activities implemented at the Intrepid offices are presented on its website.

Other examples

The tourist accommodation and attractions sub-sector, as well as events and other operations, have generated a very large number of innovative examples that illustrate opportunities for reducing GHG emissions. One reason for the broad range of initiatives is the diversity of this sub-sector and the many different sources of emissions, including buildings, other facilities and transportation. The examples provided below include the use of renewable energy sources, energy conservation and efficiency, and carbon offsetting. Accommodation examples are shown first, while good practices demonstrated by other tourism providers and facilities follow.

Company	*Initiative*	*More information*
Hidden Valley Ecolodge, Australia	Australia's first carbon neutral ecolodge, run on 100% solar power.	www.hiddenvalleycabins.com.au/
Lady Elliott Island Eco Resort, Australia	A hybrid solar diesel power station reduced carbon emissions by 70% and attracted engineers as new visitors.	Zeppel (2011)
Mashovhela Lodge, South Africa	The lodge is not on the main grid but relies on solar hot water, photovoltaic (PV), wood and other biofuel, geothermal gasification (wood gas), and a back up generator.	www.mashovhela.com
Lazy Shag Backpackers in Kaikoura, New Zealand	Installed a multi-oil clean burn boiler to supply the hostel's hot water, slashing the electricity bill. The boiler is fired with used cooking oil purchased at low cost from local businesses.	www.eeca.govt.nz/node/1417
Solscape eco retreat, Raglan, New Zealand	Installed solar water heating with a payback time of 3.5 years. Savings in electricity were 75%, or NZ$4,500 per year.	www.eeca.govt.nz/node/6338
Glazebrook Lodge, Marlborough Sounds, New Zealand	A 3.2 kW micro hydro turbine generates the lodge's electricity in a standalone power system; the Lodge also has a 27 kW back-up diesel generator.	www.eecabusiness.govt.nz/sites/all/files/glazebrook-lodge-installs-hydro-generator-june-09_1.pdf
Park Inn, Muscat, Oman	Energy efficient lighting, room sensors, low-consuming mini-bar, TV stand-by, carbon offsetting programme, as well as supporting the 'Origin Oman' campaign to use locally grown food.	www.rtd4.om/rtd/RTD4/3rd/Theme4/SushantPawar.pdf
Golden Arrow Lakeside Resort, United States	The Green Roof is a 3,400 square foot roof that creates energy savings by acting as a super insulator (i.e. warm in winter and cool in summer). It also acts as a storm water management system.	www.yourtravelchoice.org/2010/03/golden-arrow%e2%80%99s-green-business-strategies-balancing-comfort-and-conservation/

Company	*Initiative*	*More information*
Sandals Regency La Toc Golf Resort, Saint Lucia	Achieved major reductions in energy consumption (51% over five years) following its use of the EarthCheck benchmarking system.	www.earthcheck.org
Intercontinental Bora Bora (French Polynesia)	Uses cool seawater (with a pump and a cooling station) for its air conditioning system, reducing electricity consumption by 80%.	www.ichotelsgroup.com/intercontinental/en/gb/locations/borabora-thalasso
Hotel Sigiriya, Sri Lanka	Implemented a wide range of savings initiatives and reports halving the carbon footprint per guest stay.	http://64.78.36.53/Doc/Hotel_Sigiriya.pdf
Scandic Hotels, Sweden	Reduced energy consumption by 24% and water consumption by 13% (1996–2006).	www.scandic.com
Christchurch International Airport, New Zealand	In January 2008, CIAL became the first airport company in the southern hemisphere to attain certified carbon neutral status.	www.christchurchairport.co.nz/CorporateAndCommunity/TheEnvironment/
FIFA 2010 Green Goal, South Africa	Supported by a US$1 million GEF grant, this initiative supported renewable energy projects, awareness raising and carbon offsetting. One example is the use of solar energy to power twelve billboards.	www.unep.org/sport_env/presentations/GreenGoal.pdf
Manchester Airport, Great Britain	Goal: carbon neutral by 2015 for its energy and vehicle use. Investment into energy saving equipment and purchasing 100% of electricity from renewable sources; remaining shortfalls will be offset. Trials to generate energy using wind turbines.	www.manchesterairport.co.uk/manweb.nsf/Content/Environment
Ruby Falls (America's deepest commercial cave), United States	Has reduced power consumption by over 24% due to solar panels (up to 16,000 watts) and computer-controlled, high efficiency lighting in both LED and compact fluorescents.	www.rubyfalls.com/pages/Ruby-Falls-Environmental-Sustainability/
Green Island, Australia	Accreditation through: Advanced Ecotourism Certification; Leader (highest) level under Climate Action Australia's Certification Program; the Australian Government's climate change initiative Greenhouse Challenge Plus; the Planet Safe Partnership.	www.greenisland.com.au/environment/

6.5 Tourist behaviour and mitigation

Background

Awareness of climate change has increased considerably, possibly peaking in 2007 due to coinciding events such as the release of the popular film *An Inconvenient Truth*, the awarding of the Nobel Prize jointly to the Intergovernmental Panel on Climate Change and Al Gore, and the Conference of the Parties to the United Nations Framework of Climate Change Convention, held in Bali. The link between climate change and tourism was well highlighted in the media, often with a focus on the negative impact of air travel. Accordingly, a range of surveys indicated some rapid changes in tourists' awareness and perception of climate change (Brouwer *et al.*, 2008; Hares *et al.*, 2010). The 2007 Lonely Planet Annual Travellers Pulse Survey of 24,500 tourists, for example, reported that the majority of travellers were worried about carbon emissions associated with flying (Lonely Planet, 2007). Only 7 per cent of respondents said that they did not think aircraft carbon emissions were a concern.

Market research in Australia and New Zealand also indicated that tourists were concerned about the environment and the impacts of their own travel (Instinct and Reason, 2008; Tourism New Zealand, 2008). The Instinct and Reason Group, for example, surveyed 1,000 British and German long-haul travellers to Australia. They found that, overall, the German market was more sensitive to issues associated with the environment and flying than was the British market. In both markets three key (but contrasting) consumer attitudes were identified. The first was that a large proportion of both markets considered the environment to be very important and not worth putting at a risk. Second, there was a strong view that travelling long haul provided rewards associated with the enjoyment of life. Third, there was a strong tendency to either blame others, or to expect them to take action, rather than accepting any personal responsibility. To date, tourists who actually change their travel behaviour in response to climate change are very much in the minority, especially amongst experienced travellers (McKercher *et al.*, 2010). The British Department of Transport (2008) found, for example, that even though 70 per cent of British citizens believed that air travel harmed the environment, a higher proportion of people undertook multiple air trips in 2006 compared with previous years. The majority of those who travelled by air anticipated that they would travel even more in the future. This dissonance also surfaced in a number of other studies, including DEFRA's (2007) analysis of the response of British citizens to sustainable tourism and leisure and the Becken (2007) study of international tourists to New Zealand.

While concerns over climate change are unlikely to stop people from travelling, the cost of airfares might potentially have more impact (British Department of Transport, 2008; Elliott and Buckley, 2008). The literature shows that tourists, just like other consumers, react to changes in price, in relation to both destination choice and consumption of tourism products on the ground (Dwyer *et al.*, 2001; Nicolau and Mas, 2006; Brons *et al.*, 2002). Prices of tourist transport are strongly influenced by the cost of fuel. This can be affected by climate change policies that put a price on carbon emissions, for example as a result of an emissions trading scheme or a carbon tax. Several climate policy simulation studies indicate that prices would need to increase substantially (e.g. Mayor and Tol, 2007). In addition, the increase in global oil prices in 2008 provides an interesting analogue for studying consumer responses to higher fares. Fuel surcharges varied by airline, but were in the order of US$85 per long-haul flight sector (Lufthansa) to US$130 for flights between Singapore and the United States (Singapore Airlines) (Global Travel, 2008). As a consequence of higher operational costs and reduced demand, a number of airlines were close to (or in) bankruptcy and the IATA announced record losses (IATA, 2008). Unfortunately, no detailed studies (e.g. a decomposition analysis) are available that provide clear evidence on the extent to which these downturns in tourism demand in 2008/09 were due to higher transport costs or to the worsening economic situation worldwide. This highlights the complexity of factors influencing tourism demand, and the challenges researchers face in trying to model it.

So, while both the extent to which tourists translate their concerns into actual behaviour and their response to financial signals from climate policies are somewhat uncertain, a growing number of initiatives do highlight opportunities for sustainable consumption. Broadly, these can be categorized into: (i) avoiding; (ii) reducing; and (iii) offsetting carbon emissions. Avoiding carbon emissions includes decisions made by tourists, such as travelling for longer but less frequently, and visiting destinations that are closer to home. Ideally, reducing emissions relates to choices that allow the tourist to continue travelling, but with a focus on minimizing emissions. Examples include carbon efficient transport modes (e.g. trains), the consumption of low carbon services and products (e.g. locally resourced food), and energy smart behaviour (e.g. switching off unnecessary appliances in the hotel room). It is essential to provide user-friendly information on the different options, and especially their carbon footprints, so that comparisons can be made without difficulty. The third option, carbon offsetting, has become increasingly popular as it allows passengers to travel 'guilt-free'. Despite being a good option in principle, carbon offsetting has been met with increasing scepticism (e.g. Gössling *et al.*, 2007). In response to concern about the integrity of offsetting schemes, Clean Air – Cool Planet (2006) compiled a consumers' guide to carbon offsetting, containing a number of good practice principles, such as 'additionality' and 'verification' of offsets.

> Young people are living more structured lives than ever before, and as we move towards 2020, the days of carefree, spontaneous travel will disappear in favour of more organised, planned and researched trips. Students and young travellers will continue to be trailblazers, craving adventure and experiences they can share, but they will also be very aware of how their travel will contribute to the rest of their lives. Working abroad, voluntary and conservation projects, and learning something new will all be high on the travel agenda. (John Constable, managing director, STA Travel)

Case study: Providing climate information

Hospitality and food services are essential components of tourism. For many tourists, experiencing local, and exotic, food is an important part of their holiday. However, if the carbon intensity of food consumed during the holiday is larger than food prepared at home, the net carbon footprint of tourism increases (Gössling *et al.*, 2010b). The example of the Swedish fast food chain Max Burger illustrates how carbon accounting and labelling can lead to real reductions in greenhouse gas emissions related to food consumption. Max Burger claims that meals cooked at home produce around the same quantity of carbon emissions as traditional Max hamburger meals. The cornerstone of Max Burger's climate programme is the declaration of the carbon emissions for its entire product range. This information is provided on the menus. A carbon footprint assessment, undertaken by Max Burger in partnership with The Natural Step, highlighted that 70 per cent of Max Burger's total emissions are produced by the production of beef. Its total carbon footprint was estimated to be about 27,000 tonnes of CO_2 per year. To reduce this footprint the company uses only Swedish meat partners with 'high standards', although it does not explicitly explain what these standards entail. Max Burger also actively promotes vegetarian burgers. Other important mitigation initiatives include:

- the use of 100 per cent wind energy for all restaurants;
- offsetting emissions by planting trees in Africa;
- using environmentally friendly company vehicles;
- an internal environmental prize for the restaurant with the best environmental initiative;
- installing low-energy LED lighting instead of neon lighting in all restaurants; and
- recycling all electrical equipment.

A similar approach has been taken by Thai Airways, Thailand. From the beginning of 2010 the Thai inflight menu has featured carbon footprint information, starting with two Thai Signature dishes, Chicken Mussaman Curry (1.36 kg CO_2-equivalent per 250 g serving), and Green Curry 'Kiew-Wan' (1.39 kg CO_2-equivalent per 250 g serving). The calculations use life cycle assessment. Carbon footprint labelling is part of Thai Airways' participation in the 'Carbon Footprint Technical Cooperation Project for Thai Products', in cooperation with the Thai Greenhouse Gas Management Organization and the National Metal and Materials Technology Centre. This project utilizes the Thai National NCI Database and life cycle assessment to increase production capabilities and boost competitiveness amongst Thai industries, including assessment of greenhouse gases emitted from these pilot products. The aim is to lower energy consumption and maximize product efficiency. Providing such information to customers allows them to participate in carbon management as well as make informed choices.

Case study: carbon offsetting in tourism

Travellers wanting to offset their carbon emissions can choose amongst different offset providers, typically by using online services. A recent comparison of five providers for a flight from New York to the Caribbean showed that the costs to offset CO_2 emissions differ vastly (Table 6.6). To some extent this reflects different assumptions (e.g. including a factor to account for climate impacts at higher altitude) as well as the effectiveness and integrity of the offsetting programme.

Atmosfair, a Germany-based organization, is consistently listed amongst the most credible carbon offset providers (e.g. Clean Air – Cool Planet, 2006). Atmosfair was founded in 2005 and is a joint initiative of 'forum anders reisen' and the environment and development organization Germanwatch. It is also supported by the German Federal Environment Agency. The Atmosfair website includes climate change information (e.g. about the climate impact of flying), provides the opportunity to make a donation or to buy carbon offsets for a particular trip and describes which tour operators and travel agencies cooperate with Atmosfair. Most recently, Atmosfair released an Airline Index in which airlines around the world are ranked in terms of their climate impact, taking into account such factors as fleet composition and age, and climate mitigation measures.

The Atmosfair offsetting projects are all located in developing countries and meet the Gold Standard criteria – a quality label for offsetting projects that explicitly demands the principle of additionality. This means that the emissions reductions used as offsets are not 'business as usual' – in other words, the reductions would have happened even if offsetting had not been used. Atmosfair's projects include, amongst others, biogas production from cow dung in India (Plate 6.11), the use of solar mirrors for cooking in India, a micro-dam project for rural electricity generation in Honduras, using energy from peanut residues in Burkina Faso and the treatment of wastes from palm oil production in Thailand. All emission reductions are Certified Emission Reductions under the UNFCCC. Detailed reports and more information can be found at www.atmosfair.de.

TABLE 6.6 Comparison of offset providers from New York to the Caribbean. (Simpson *et al.*, 2008a)

Offset company	*Tonnes of CO_2*	*Cost/tonne of CO_2*	*Cost of offset (US$)*
Atmosfair	1.42	30.97	43.98
The Carbon Neutral Company	0.60	15.3–28.55	9.18–17.13
Climate Care	0.55	15.80	8.69
My Climate	1.16	15.30	17.75
Climate Trust (via Carbon Counter)	1.84	12.00	22.08

PLATE 6.11 Carbon offsetting project by Atmosfair: Cow dung is fermented in underground biogas digesters, producing an almost odourless and high-energy content gas. (Atmosfair Annual Report 2009, www.atmosfair.de/fileadmin/user_upload/Medien/Jahresbericht_2009_EN_Website.pdf)

Other examples of mitigation related to tourist behaviour

An important step to changing tourist behaviour is the provision of relevant information. The compilation of examples below highlights a range of information initiatives and campaigns that help tourists make carbon efficient decisions. Other examples relate to a tourist's contribution to mitigation through ecosystem protection and carbon sequestration.

Company	*Initiative*	*More information*
SnowCarbon ski resorts by train, Great Britain	This is an independent website dedicated to helping skiers and snowboarders reach ski resorts by train from the UK. The website is an official booking site for trains and also provides comprehensive ski packages.	www.snowcarbon.co.uk

Company	*Initiative*	*More information*
Rough Guide Climate Change	Comprehensive 'guidebook' on climate change to inform travellers.	www.roughguides.com/website/shop/products/Climate-Change.aspx
Carbon Friendly Flight Search	This online tool provides information on flights (linking with existing flight search engines such as Opodo) and a carbon ranking based on routing and fleet efficiencies.	www.globaltravelmarket.co.uk/flights-engine/?client=carbonResponsible
Enjoy England	England's official tourism website, with dedicated sites to carbon footprint reduction, responsible tourism and green accommodation.	www.enjoyengland.com
Green Passport	UNEP initiative to raise tourists' awareness of their potential to contribute to sustainable development by making responsible holiday choices. Paraty, a city in Brazil, is identified as the first tourism destination to implement the communication materials.	www.unep.fr/scp/tourism/activities/taskforce/pdf/fact%20sheets%20pdf/-green_passport.pdf
Tourism Queensland, Australia	Facilitates project involvement for volunteer tourists, including activities related to tree planting and nature conservation.	www.seagrasswatch.org/Info_centre/Publications/Volunteer_Tourism_Conservation_Programs.pdf
Amazon Lodge, Ecuador	Tourists participate in a reforestation project in the Amazon forest. Volunteers stay between two to four weeks.	www.volunteerabroadworld.com/volunteer/4757-South%20America%20Inside.htm
Adventure Expeditions, Peru	As part of an adventure-multisport holiday, tourists contribute through manual labour to reforestation (e.g. building a fence).	www.adventureexpeditions.net/trip_daily_overview.php?trip_id=trip_4c1d2918f2460867882

6.6 Conclusion

This chapter demonstrates that there are many excellent examples of how the tourism sector has become involved in climate change mitigation. Many tourist destinations, and also tourism-dependent countries, are showing increasing interest in reducing their GHG emissions and improving their 'carbon footprints', some even to the extent of wanting to become carbon neutral. Some countries, including Australia, have dedicated considerable resources in order to better understand and manage the impacts their tourism sector has on the global climate.

Similarly, more and more tourism businesses are recognizing the co-benefits of saving energy and reducing GHG emissions. The high oil prices in 2008, and again in 2011, add additional incentives to invest in new technologies, make managerial improvements and change behaviours. However, higher oil prices (and therefore increasing costs, e.g. for transportation) may lead to a reduced uptake of voluntary carbon offsetting. Importantly, no research has been conducted on this potential trade-off. It is still uncertain whether tourists truly respond to the environmental efforts by tourism operators. We also do not know the full extent to which tourists are willing to engage in climate change mitigation. The available tourism literature on these topics indicates high levels of awareness and intent to contribute, but low levels of practical action. Notwithstanding, a number of informative examples of consumer information and behaviour to reduce emissions have been discussed in this chapter. As also summarized by Scott and Becken (2010), the collective findings raise the important question as to whether the challenge of climate change can be left to the market forces of demand and supply alone, since tourists might lack the commitment to be 'ambassadors' for climate change (see also Eijgelaar *et al.*, 2010).

A growing number of tourism businesses and other stakeholders include climate change in their publicity and activities. The selections presented in this chapter are only a small sampling of all the mitigation activities underway in tourism. Case studies and examples were chosen to highlight genuine, and often innovative, efforts to reduce GHG emissions. While making the selections it became evident that often it is not easy to distinguish technically credible mitigation actions from those that are more driven by marketing motivations, thus representing 'window dressing'. Some sub-sectors of tourism are more astute than others when emphasizing their proactivity in relation to addressing climate change. The aviation sector is a good example, with many airlines becoming heavily involved in both addressing climate change and talking about it. Many airlines have web pages dedicated to climate change and document relevant activities in their annual reports, often highlighting the 'uniqueness' of their initiatives. On the other hand, some airlines have failed to actively engage in climate change mitigation and make use of fuel reduction opportunities.

It is important to comment on the geographical spread of appropriate examples of climate change mitigation in tourism. While it was our intention to provide informative examples from around the world, in the end examples from Europe, Australia and New Zealand dominate. It was extremely difficult to find case studies that highlighted good practices in Africa, Asia and South America. Even such examples from North America are few in number. To some extent these differences reflect the contrasting policy environments, as well as cultural contexts. The European Union, for example, is generally very much committed to developing and implementing strong environmental policies, including those with relevance to tourism operations and destinations within Europe. The engagement by Australia and New Zealand may partly reflect the media spotlight on these two countries in the last few years in relation to long-distance travel and aviation emissions. Since 2007, both countries have become quite concerned that negative publicity around carbon footprints could seriously undermine their tourism industries. The initiatives implemented since then may also relate to the pristine (e.g. 'clean and green' in the case of New Zealand) image that both Australia and New Zealand enjoy, and that underpins their marketing campaigns. Similarities between the examples used in the preceding Mitigation Policy chapter and the current chapter on Mitigation Practice highlight the benefits of there being a strong enabling environment if mitigation efforts are to be effective in terms of actual reductions in GHG emissions. This includes ensuring that the industry is well informed, and that efforts are supported rather than impeded by institutional arrangements and regulatory measures.

7

ASSESSING PERFORMANCE, AND LEARNING

7.1 Introduction

It is important to know whether our responses to climate change are proving to be effective, and to reinforce the successful actions and modify or terminate those that are not achieving the desired results. Commonly this process of assessing performance is called 'monitoring and evaluation' (Figure 7.1). The monitoring and evaluation of adaptation and mitigation covers not only specific, on-the-ground interventions, such as those described in Chapters 4 and 6, but also the policy and planning initiatives, including those described in Chapters 3 and 5. While in many cases evaluating climate change interventions is challenging, it is important to do so. This is in part because evidence of the effectiveness of current spending on responses to climate change is essential at all business, national and global levels. For example, such information is required if current proposals to increase the financing of climate change mitigation and adaptation by hundreds of billions of dollars a year are to be treated seriously and receive increased support (Prowse and Snilstveit, 2009).

In the following sections we aim to provide practical, tourism-focused information and guidance related to the monitoring and evaluation of mitigation and adaptation interventions. Progress in the area of evaluation of tourism initiatives in relation to climate change will increase robustness and credibility of both mitigation and adaptation. Advertisement campaigns of (claimed) low carbon products will experience greater scrutiny, but will hopefully also gain credibility (Plate 7.1).

To date, performance assessments have tended to be mutually exclusive, focusing on either mitigation or adaptation. However, we also consider an important link between these two action areas, as also shown in Figure 7.1. Financial resources generated by taxing the GHG emissions by the tourism sector, and especially commercial aircraft emissions, are being identified as a way to fund adaptation, not only for the tourism sector, but more widely (Baker, 2011). Finally in this chapter we

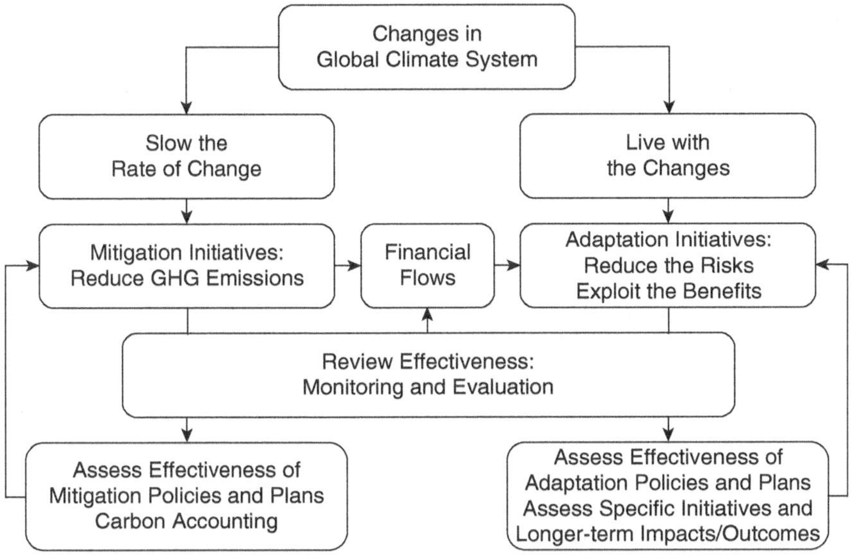

FIGURE 7.1 The role of monitoring and evaluation in the context of mitigation and adaptation, and of mitigation as a source of financial resources to support adaptation.

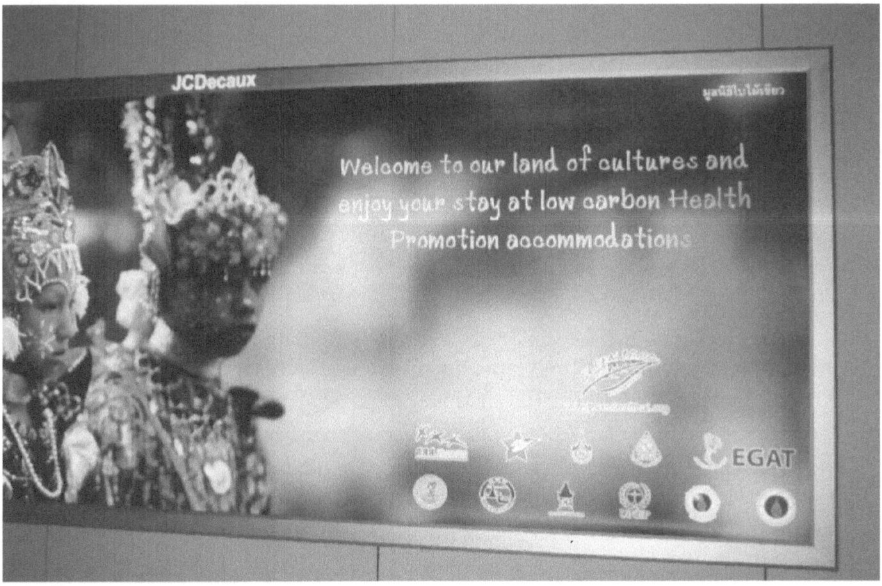

PLATE 7.1 Promotion of 'low carbon health promotion' accommodation at Bangkok airport. (S. Becken)

highlight the learning opportunities that come from monitoring and evaluating tourism-related mitigation and adaptation initiatives at all levels, from policy making to technical interventions. The resulting understanding is used not only to improve adaptation and mitigation performance by the tourism sector (Figure 1.3) but can also be useful in understanding the limits to adaptation and the implications of setting mitigation targets through international negotiation processes (IPCC, 2007a).

7.2 Assessing the performance of mitigation policies and plans for the tourism sector

In the context of mitigation, accounting for carbon emissions is the main approach used to identify both the opportunities to reduce GHG emissions and the extent to which current interventions are achieving the planned reductions. Establishing a baseline (and benchmark) for emissions, for example through emission coefficients (Figure 7.2), is an important step in GHG monitoring and reduction programmes. There are very few analyses that systematically assess the effectiveness of different mitigation policies. Friedrich *et al.*'s (2011) multi-criteria analysis of 12 policy instruments in the Trinidad and Tobago power sector could be used as guidance for future tourism policy analyses. The results of their study indicate that the coexistence of multiple policies is beneficial, although the provision of subsidies for energy saving technologies, an industry-wide carbon trading scheme and the establishment of a feed-in tariff for renewable energy sources emerged as the most effective policies.

Carbon accounting

While policy analyses are scant, the measuring of GHG emissions (often including non-carbon GHG gases through the use of CO_2-equivalents) has become increasingly important and common for tourism businesses and organizations. A vast number of tools are now available to measure or estimate emissions. Examples include:

- the Travel Analytics Airline Carbon Emissions Calculator;[1]
- ICAO's Carbon Calculator for air travel;[2]
- the Sustainable Tourism CRC Carbon Calculator for tourists in Australia;[3]
- the CarboNZero Travel and Tourism Calculator for tourists in New Zealand;[4]
- the EarthCheck Carbon Calculator for Tourism Businesses;[5]
- the Carbon Footprint™ Calculators for businesses, individuals and car travel;[6] and
- the Hotelenergysolutions.net Toolkit and carbon calculator for European hotels.[7]

FIGURE 7.2 The carbon intensity of travel options – a webcomic. Reprinted with permission from Ecometrica (www.ecometrica.co.uk). (Based on the paper 'Carbon Intensity of Travel Options: 2011 Update')

The earlier book of Becken and Hay (2007) provided a detailed description of GHG accounting methods of relevance to the tourism sector, including those related to top-down and bottom-up approaches. The different types of fuels (e.g. oil, gas and coal) were discussed and illustrative emission factors were provided. Most countries publish emission factors that they use in their national GHG inventories. These will reflect the types of energy sources and uses within the country. In the absence of such national coefficients, the IPCC maintains an Emission Factor Database. This is a recognized library of background information, emission factors and other parameters that can be used for estimating both GHG emissions and removals (IPCC, no date).

Accounting frameworks are also important, as these define the boundaries of the system. The most common (non-tourism) framework is the Greenhouse Gas Protocol. This was developed in a partnership between the World Resources Institute and the World Business Council for Sustainable Development, with contributions from businesses, governments and environmental groups around the world. In essence, the GHG Protocol offers guidance, standards and tools for calculating and managing emissions associated with a business. For service sector companies, emissions are categorized into three scopes (Figure 7.3). Scope 1 relates to all the emissions directly generated at the business (e.g. the hotel); Scope 2 includes emissions associated with purchased energy; and Scope 3 considers all other indirect emissions, including those related to business trips (e.g. air travel).

More recently, the International Tourism Partnership (ITP) and the WTTC are collaborating on an initiative to harmonize carbon calculation efforts by hotels around the world. This includes agreeing on a standardized methodology and metrics informed by the GHG Protocol standards. The goal is to achieve agreement on a single methodology for calculating carbon footprints, as well as consistent approaches for communicating emissions. Emissions caused by tourists are typically

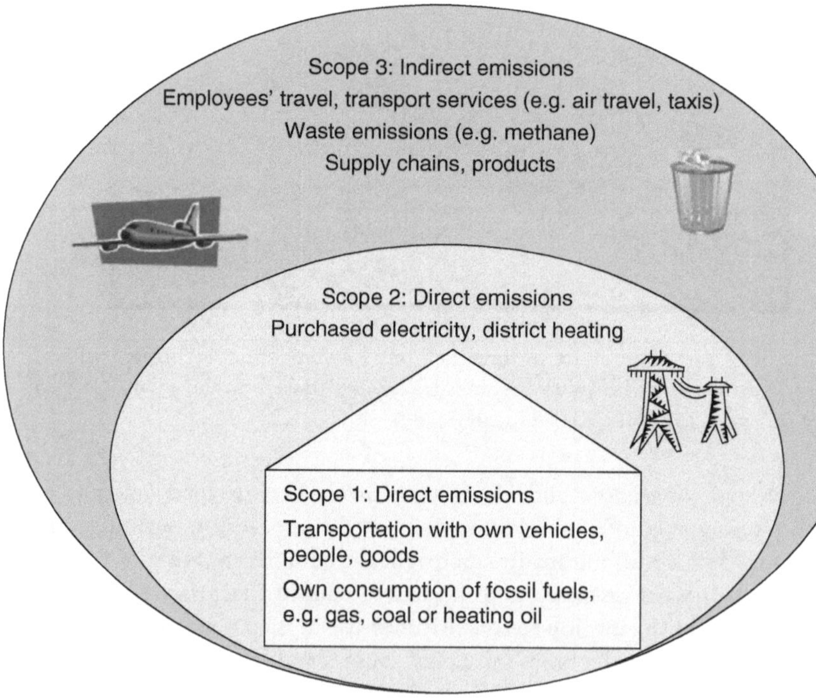

FIGURE 7.3 The three scopes of the Greenhouse Gas Protocol in the context of tourism.

not included in business inventories and fall within the range of individual responsibility by tourists (see Chapter 6 for comments on the environmental awareness and behaviour of tourists).

Another framework is 'PAS 2050 – Assessing the life cycle greenhouse gas emissions of goods and services', developed by the British National Standards Body. PAS 2050 considers the entire life cycle of a product, for example, an apple. This standard has not yet been applied to tourism, but it could be used to compare the emissions associated with different types of tourism end-products, such as a night in a hotel or a holiday package. Kuo and Chen (2009) provide an example of the application of generic (rather than part of PAS 2050) life cycle assessment to tourism in Taiwan.

A product-based approach could potentially be significant for tourism, given its high dependence on intermediary inputs (e.g. food supplies, construction and transportation services). For example, a recent study found that the energy embodied in one litre of water (in a plastic bottle) consumed in Los Angeles can range from 1.4 MJ when produced locally, to 4.0 MJ when shipped from Fiji and 5.8 MJ when imported from Evian in France (Gleick and Cooley, 2009). Tourism often occurs in remote areas, such as islands. As a result, the indirect carbon footprint is often high (Becken, 2004).

Defining the baseline

Defining the baseline is important, and can form part of wider environmental reporting, for example through CSR reports. While preparing such reports is not widespread amongst smaller tourism companies as yet, larger corporations are increasingly engaging in CSR, for example the airline industry (Cowper-Smith and de Grosbois, 2011). A 2009 study on CSR by airlines found that 14 airlines (out of 41) had an annual CSR report available. Emission reductions dominated as a theme, but social sustainability also featured. Companies use different measurements, making it difficult to compare airlines. The research also revealed that most reports state goals and commitments, while much less information is available on how these will be achieved through tangible actions. Thus, while some progress is being made in relation to carbon accounting and defining baselines, there is still ample opportunity for improvement and for carbon accounting to be used more widely in the tourism sector.

Increasingly, destination marketing organizations are undertaking surveys to monitor not only tourists' satisfaction, but also their environmental awareness and perceptions of the destination and individual aspects of it (e.g. transport, accommodation and activities). Examples include surveys conducted by Tourism New Zealand and Tourism Australia. In order to measure change in such attributes as satisfaction, awareness and perceptions it is important that an initial survey be conducted in such a way that a comprehensive baseline is established, against which the changes can be measured.

Case study: A national carbon account for tourism

Accounting for all GHG emission at a national level, or even for CO_2 alone, can follow a bottom-up or top-down approach (Becken and Patterson, 2006). Both approaches were recently demonstrated for Switzerland by Perch-Nielsen *et al.* (2010). The advantage of the top-down approach is compatibility with carbon accounting by other sectors, as well as with national accounts. In their paper on estimating the carbon footprint of Australian tourism, Dwyer *et al.* (2010) provide methodological details for two slightly different top-down approaches: a 'production approach' and an 'expenditure approach'. When the Tourism Satellite Account is used they found that, depending on the approach, in 2003–04 tourism contributed between 3.9 per cent and 5.3 per cent to Australia's total industrial GHG emissions. The rationale for each approach is explained and a comparative overview is presented in Table 7.1. The analysis by Dwyer *et al.* (2010) highlights how the amounts of GHG emissions from tourism depend on the accounting framework. These and similar findings highlight that a common approach to carbon accounting for tourism is highly desirable, if only for reasons of comparability.

TABLE 7.1 National GHG account for tourism, Australia. (After Dwyer *et al.*, 2010)

Production based approach	*Expenditure based approach*
Includes: • GHG emissions directly produced by tourism industries; • GHG emissions from Australian-based airlines; and • GHG emissions from imports used as inputs in producing goods/services for sale to the Australian tourism industry.	**Includes:** • GHG emissions associated with expenditures in Australia by foreign and domestic tourists; • GHG emissions from air travel on Australian and non-Australian-based airlines; and • GHG emissions associated with expenditure by outbound Australian residents within Australia prior to or following flights.
Excludes: • GHG emissions of non-Australian-based airlines; and • GHG emissions from production of imports directly purchased by visitors.	**Excludes:** • GHG emissions associated with expenditure on outbound airfares; and • GHG emissions associated with expenditure by Australian outbound travellers on goods/services outside Australia.

Case study: A sector approach to GHG accounting and mitigation

The DEHOGA (Deutscher Hotel und Gaststättenverband) has developed an online system to help members (hotels in Germany) reduce their energy use and GHG emissions. The initiative is called Energiekampagne Gastgewerbe.[8] It provides a large range of technical and organizational guidance for hotels and restaurants. A cornerstone of the system is a 'personal energy savings account' which can be established online and which serves as an ongoing repository for records of the member's energy consumption and emissions. Based on the data provided, personalized information will be provided to the business, including recommendations for specific mitigation measures, comparisons with other businesses and financial savings potentials. Participation is simple, and involves three steps:

1. Registration online and provision of some basic data (taking about 20 minutes to enter).
2. Completion of four energy questionnaires on: (i) the business; (ii) the building; (iii) technologies employed; and (iv) management. This step includes providing data on energy costs (power, heating fuel, gas, district heating and renewable energy sources), ideally for a whole calendar year (often used to cover different seasons).
3. Analysis of data. The online programme automatically calculates standardized energy parameters and individualized and practical savings suggestions (especially in the areas of heating, lighting, energy management, cooling and maintenance).

The programme also provides a benchmark and compares the participant's business with other operators in similar categories. The Energiekampagne allows monitoring over time, for both the business itself and in comparison to others and the sector benchmark. The results are summarized in a report which can be used to communicate improvements to guests and other stakeholders. Users can also learn from reports of (currently 54) case studies. There is no charge for using the tool.

Case study: Measuring the emissions by a destination

While tourist destinations are increasingly eager to become involved in climate change mitigation, and maybe even brand themselves as carbon neutral (see Chapter 5), there are no practical examples of a destination that has actually undertaken a detailed inventory for this purpose. A model of how to approach this task has been presented by Kelly and Williams (2007) for the case of Whistler, British Columbia, Canada – an important destination for recreation, including skiing.

Whistler has a history of approaching energy, air quality and GHG emissions in an integrated way. In 2004 the municipality adopted an Integrated Energy Plan, consistent with the resort's emphasis on sustainability. The Plan was community driven and a Whistler 2020 taskforce was established for implementation. While the Plan, and engagement with the community, are a success story, the actual achievements in terms of GHG reductions are behind schedule. One of the main (and unresolved) problems is transportation to and from Whistler. The municipal council and Tourism Whistler are now collaborating to provide carbon calculators and offsetting options as part of their booking systems for tourists (Commission for Environmental Cooperation, no date).

A 'Resort destination energy consumption and emission model' illustrates how tourism can be included (or segregated) from the municipality's energy use and GHG emissions (Kelly and Williams, 2007). The model includes three key drivers of emissions: (i) energy use at the destination, for example for buildings, infrastructure and local transportation; (ii) commuting by employees to and from Whistler; and (iii) travel by tourists to and from Whistler. The model also makes allowance for policies and characteristics of the energy system. The system is composed of eight components (Table 7.2) that are inter-linked through multiple relationships and feedbacks.

The study also demonstrates how, through scenario analysis, changes in land-use policies, building design and infrastructure investments can be shown to have profound impacts on a destination's emission profile. Planning strategies can also influence visitor perceptions and decision making, for example as a result of the development of an attractive destination brand. The local population is another important element of the emission model as it provides employees to work at the resort, and as a result contributes to tourism emissions (e.g. through commuting).

TABLE 7.2 Elements of the energy system that determine energy consumption and GHG emissions in Whistler, Canada. (After Kelly and Williams, 2007)

System elements	*Detail*
Planning strategies	Land use allocation, building design, transportation, infrastructure
Local population	Permanent, seasonal, non-working
Employee commuting	Single occupancy, car pool, train, other
Intra-community transport	Personal vehicle, community transport, municipal vehicles
Visitors	Day, visiting friends, paid accommodation, second home
Travel to/from Whistler	Personal vehicle, bus, train, plane
Buildings (residential, commercial, institutional)	Single family, apartment, hotels, public buildings, service, wholesale, storage, etc.
Energy supply systems	Electricity/grid, wood, gas, local renewables, etc.

Case study: Carbon accounting at the business level

The AUS$125 million Emirates Wolgan Valley Resort in Australia is being built on the same principles as Emirates' acclaimed Al Maha Desert Resort and Spa. The plan is to make it compatible with the surrounding environment, and to support conservation projects. The resort operates on about 4,000 acres of land, which are designated as a wildlife reserve. In addition to the conservation initiatives, the resort is undergoing carbon-neutral certification through CarboNZero in New Zealand. To date the resort has:

- measured its greenhouse gas emissions using ISO 14064–1:2006 compliant methodologies;

- committed to managing and reducing its emissions; and
- offset the remaining unavoidable emissions by acquiring the requisite number of verified carbon credits (see Chapter 6 on carbon offsetting).

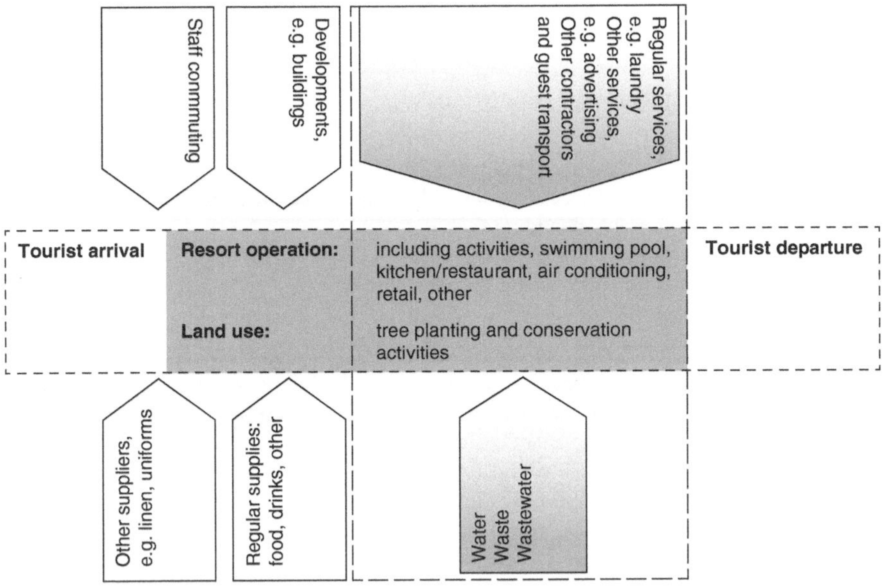

FIGURE 7.4 Diagram of the supply chains of Emirates Wolgan Valley Resort in Australia, showing sources of emissions, and highlighting those units that are included in the GHG inventory. (After CarboNZero at www.carbonzero.co.nz/documents/disclosure_wolgan_0910.pdf)

As noted above, an important part of carbon accounting is determining appropriate boundaries. Figure 7.4 shows the scope that was selected for the Emirates Wolgan Valley Resort. Unshaded units were excluded from the certification while the shaded ones were included. Significantly, although guest travel to and from the resort is likely to be a substantial contributor to GHG emissions, it is not included in the resort's carbon account. This is a common issue with 'carbon footprints' of tourism businesses and a major reason why actual emissions are often underestimated.

Further examples of carbon accounting initiatives

There are many examples worldwide that highlight progress in carbon accounting in the tourism sector. Several initiatives related to carbon certification, inventories, offsetting and management are summarized overleaf.

Company	*Initiative*	*More information*
Airport Carbon Accreditation Europe	European carbon standard for airports. The programme assesses the efforts of airports to manage and reduce their carbon emissions with four levels of award: 'Mapping', 'Reduction', 'Optimisation' and 'Neutrality'.	www. airportcarbonaccreditation.org/
GRI for Airport Operators	The Global Reporting Initiative (GRI) developed guidance for airports to account for a wide range of impacts, including GHG emissions. The goal is to enable operators to develop Sustainability Reports.	www.globalreporting.org/ReportingFramework/SectorSupplements/Airports/
World Expeditions	With the help of Planetair, World Expeditions estimated the carbon footprint of each holiday package, including internal flights, road transport, accommodation and food. The equivalent offset amount is included in the trip cost. An air travel emissions calculator on the company's website aims to encourage tourists to offset emissions.	www.worldexpeditions.com/nz/index.php?section=responsible_travel
Spirit of Japan Travel, Japan	All the carbon emissions associated with accommodation, office energy use, and transportation provided for travellers and guides are calculated and offset by investing a portion of company profit into non-profit environmental organizations, and into renewable energy initiatives.	www.eco-tourism.jp/sustainability.php
Green Leaf Certification, Thailand	Certification and ecolabelling programme in Thailand to encourage measuring and reducing the consumption of energy and other resources.	www.hotelthailand.com/greenleaf/
Hilton Environmental Reporting (HER)	Collects information on energy use per guest night and per square metre, and calculates kilograms of CO_2 emitted per guest night. The Hilton chain has data for each hotel since 2005, and uses this information to improve its energy performance.	Zientara and Bohdanowicz (2010)

Company	*Initiative*	*More information*
Green Engage, Intercontinental Hotels Group	900 hotels are using an online emission measuring and monitoring tool and have achieved savings of 10%.	www.ihgplc.com/index.asp?pageid=742
Sustainable Tourism Australia (Ecotourism Australia)	The Climate Action Certification Program is dedicated to ranking efforts to reduce carbon emissions. It is designed for all sectors of the tourism industry. This certification scheme intends to be a user-friendly, online questionnaire with online instant help desk. It also seeks to actively involve government and industry partners in the further development and marketing of the scheme.	www.sustainabletourismaustralia.com/

7.3 Assessing the performance of adaptation policies and plans for the tourism sector

The monitoring and evaluation of adaptation is still in its infancy. Moreover, relative to monitoring and evaluating mitigation, it is somewhat more complex and challenging, both conceptually and in practice (Sanahuja, 2011). An important goal for a monitoring and evaluation system for adaptation is to identify the aspects that are working, those that are not working, and the reasons why, as well as providing mechanisms to adjust the adaptation process accordingly (GEF, 2008). However, Sanahuja (2011) argues that, while effective adaptation is an important part of adapting well, the purpose of evaluation should not be about determining the success or failure of adaptation interventions, but rather about learning from the process.

Monitoring and evaluation also assists practitioners in managing their work in the context of the uncertainty surrounding climate change impacts (GIZ and WRI, 2011). Strategies designed to guide complex and long-term political processes should not be considered static, never-to-be-repeated. This holds particularly true for climate change adaptation policies and plans, given that they have to accommodate future projections, factual uncertainties, intricate interactions and political realities. Thus it is unwise to assume that a strategy or plan will remain consistent with the current scientific understanding and will continue to represent an appropriate, implementable, accepted, effective and lasting solution to all relevant climate-related risks and opportunities. Jopp *et al.* (2010) have made specific reference to monitoring and evaluating in the context of implementing a framework for regional destination adaptation to climate change. They note that, as climate change often presents long-term impacts such as sea-level rise, or deals with infrequent events such as severe storms, droughts or damaging wildfire, the evaluation of adaptation strategies is intrinsically problematic. The evaluation of adaptation must occur over the long term, with the situation being continually monitored to determine if the intended benefit has eventuated. The fact that an adaptation intervention has not yet delivered the expected amelioration of risks, or the benefits, does not necessarily mean the adaptation was inappropriate.

For these reasons, adaptation policy making and planning must be designed as a continuous and flexible process and subjected to periodic review. Based on Swart *et al.* (2009), the following are presented as useful – if not indispensable – elements of successful adaptation policies and plans, allowing for dynamic improvement and facilitating successful implementation:

- a review mechanism/institution ensuring that new scientific evidence and experiences from the implementation process will be evaluated by a competent body and reflected in an updated policy or plan on a regular basis;
- a monitoring programme that allocates responsibilities and covers the observation of actual variations and changes in the climate, the resulting impacts and progress and shortcomings in implementing the policy or plan;
- specific objectives, indicators and targets that make it possible to clearly identify the progress made and the actions that are still required;
- reporting obligations that compel key actors to provide an account of the measures they have taken, and the extent to which they have accomplished the aims and/or tasks set out by the strategy;
- adaptation planning instruments requiring stakeholders to precisely determine and regularly evaluate the ongoing need for their actions, as well as the adaptation aims and programme of measures;
- binding legal obligations if voluntary action is hindered by a lack of incentives; and
- incentives and compensation measures that allow proactive enforcement where contrary interests must be overcome.

As noted in Chapter 3, few policies and plans for adaptation in the tourism sector include these desirable attributes of a monitoring and evaluation strategy. Thus the following sections will focus more on what could, and desirably should, be included in a tourism adaptation policy, or supporting implementation plan, with regard to monitoring and evaluation.

Defining the baseline for adaptation interventions

A particular challenge to evaluating the effectiveness of adaptation initiatives is that any variation in resilience to climate change must be measured against a moving baseline of conditions. The climate baseline is not the only moving baseline.

There is also constant change in terms of socioeconomic conditions, infrastructure, demographics, technologies, political context and other conditions. All these changes reduce or increase resilience to climate change and must be taken into account if the measured outcomes and impacts are to be attributed to the adaptation initiative with an appropriate level of credibility (GEF, 2008).

The Great Barrier Reef Tourism Climate Change Action Strategy (2009–2012) (GBRMPA, 2009) includes undertaking a survey of operators to assess the level of knowledge held by operators, identifying further information requirements operators have in relation to climate change, assessing the existing information for accessibility and usability, and identifying willing participants for upcoming climate change pilot programmes. The survey is also intended to provide a baseline for the future measurement of the effectiveness of initiatives.

Assessing the performance of specific adaptation initiatives in the tourism sector

As late as 2008, and with respect to climate change adaptation initiatives in general, Hedger *et al.* (2008) claimed that very few evaluations had been undertaken. The situation has changed substantially since then, with almost all adaptation projects funded by the Global Environment Facility (GEF), donors and other development partners including monitoring and evaluation as an integral part of project design, implementation and closure. However, the monitoring and evaluation of adaptation continues to be one of the weakest areas of adaptation practice at large.

It is important to distinguish between monitoring and evaluating progress in implementing adaptation interventions, and measuring the longer-term effectiveness of adaptation policies and specific interventions (Sanahuja, 2011). The former is likely to be much easier to measure. But there is no guarantee that delivery of an intervention will also deliver effective adaptation (Hedger *et al.*, 2008). Of those evaluations carried out to date, most have been undertaken during or immediately after implementation, with only a few focusing on evaluating the impacts of interventions subsequent to project completion (McDevitt, 2010).

For adaptation initiatives in the tourism sector, this distinction between monitoring and evaluating the adaptation process itself, and evaluating its shorter-term outcomes and longer-term impacts, can be clarified by way of examples of relevant indicators (Figure 7.5). Indicators help simplify, quantify, standardize and communicate the complex and often disparate data and other information that underpins monitoring and evaluation (UNFCCC, 2010). The hypothetical examples shown in Figure 7.5 are based on the Snowbowl Ski Resort in Arizona, United States (see Chapter 4). Indicators are typically evaluated against the baseline (see preceding section) and performance targets. For example, one of the major targets in the Upgrade Plan for the Snowbowl Ski Resort is to have snowmaking capability on 95 per cent of the ski trails.

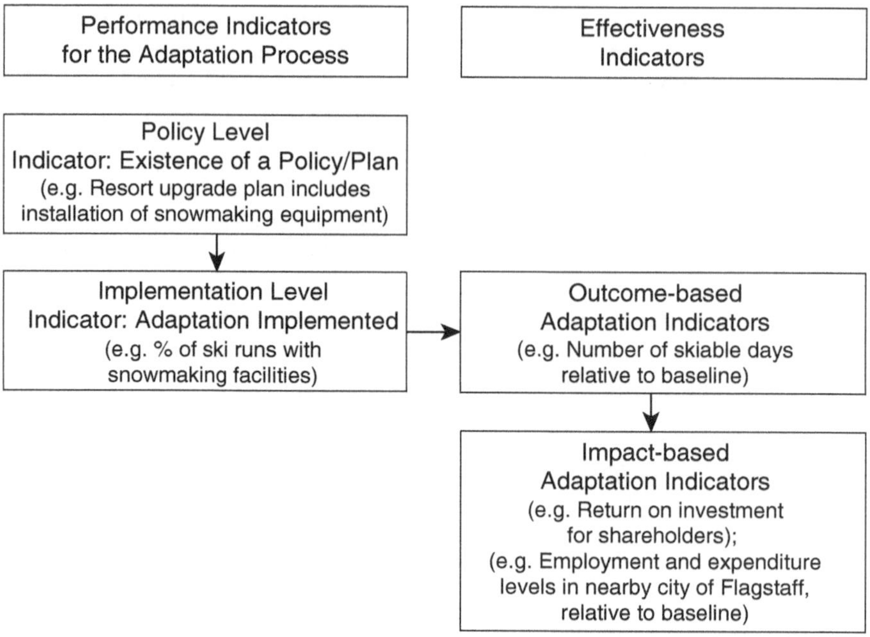

FIGURE 7.5 The two main categories of indicators for monitoring and evaluating adaptation to climate change, with examples based on Snowbowl Ski Resort in Arizona, US.

Case study: Monitoring and evaluation of adaptation by the Caribbean tourism sector

Indicators for assessing the sustainability of tourism destinations, including those related to climate change, have been identified and discussed (UNWTO, 2004), while Simpson *et al.* (2008a) have developed a set of climate change related indicators for destinations in the Caribbean (Table 7.3). These cover both adaptation and mitigation, but emphasize the former.

In 2005, the Caribbean Disaster Emergency Management Agency endorsed the Enhanced Comprehensive Disaster Management Strategy and Programming Framework. Its objective is to strengthen national and community-level capacity for reduction, preparedness and coordinated response and recovery to natural and technological hazards and the effects of climate change.

Subsequently a 'Regional Disaster Risk Management Strategy and Action Plan for the Tourism Sector' was prepared. The aim of the Strategy is to build a culture of natural hazard risk reduction within the Caribbean tourism sector in order to minimize potential deaths and injuries, loss of property, livelihoods and economic activity, and damage to the environment, through strengthening capacity for reduction, preparedness, response and recovery. Stakeholders recognized the need for a regional mechanism for monitoring, evaluating and reporting the

TABLE 7.3 Proposed climate change related indicators for sustainable tourism destinations in the Caribbean. (Simpson *et al.*, 2008a)

National standards exist for the construction of new buildings to be set back from the shoreline.
A climate change risk assessment for tourism industry has been completed.
An assessment of destination's adaptive capacity to climate change has been completed.
A system to measure and monitor carbon emissions in destination is being used.
Percentage of energy consumed in the destination from renewable sources.
Percentage of beaches where erosion is monitored at least annually.
Percentage of coastline with visible signs of erosion.
Effective protection measures in place in vulnerable areas.

results of the implementation of the Strategy and Action Plan, as well as for climate change adaptation. A project to develop a regional monitoring and evaluation system for disaster risk management and climate change adaptation in the Caribbean tourism sector was recently launched. Figure 7.6 provides an overview of the project activities and intended outcomes and impact.

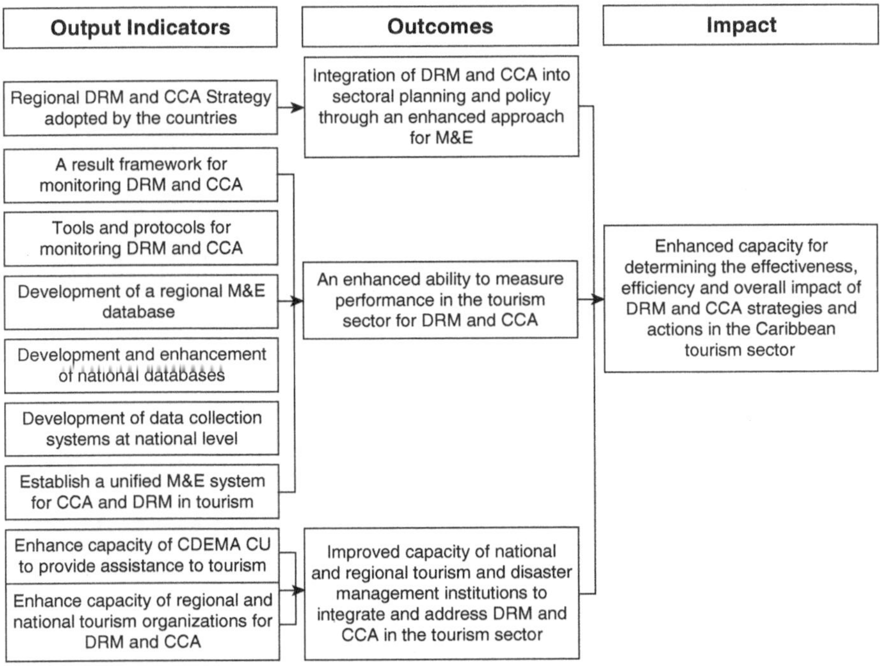

FIGURE 7.6 Overview of the activities and intended outcomes and impacts of a regional monitoring and evaluation (M&E) system for disaster risk management (DRM) climate change adaptation Caribbean Disaster Emergency Management Agency (CDEMA) and Coordination Unit (CU) in the Caribbean tourism sector. (After Persad, 2011)

Case study: Monitoring and evaluation of adaptation by the tourism sector of the Maldives

The 'Increasing Climate Change Resilience of Maldives through Adaptation in the Tourism Sector' project that has commenced recently will provide the tourism sector in the Maldives with policy and regulatory support and technical skills and knowledge. The objective of the project is to increase its adaptive capacity to respond to the impact of climate change and invest in appropriate adaptation measures that make sense in development terms, whether or not a specific climate threat actually materializes in the future. The main indicator of project success will be the tourism sector being prepared to address climate change risks, resulting in tangible private-sector investments in adaptation, environmental management and risk financing measures. The project will establish at least ten new investment projects to modify operational infrastructure and increase resilience to the impact of climate change, as well as ten community-based adaptation projects involving tourism-associated communities and operators (Plate 7.2).

Table 7.4 presents the part of the project results framework relating to one of the outcomes. It shows the relationship between the outcome, the project baseline, and the performance indicators, targets and means of verification.

PLATE 7.2 Overwater bungalows are an important part of the brand of tourism in the Maldives. (J. E. Hay)

TABLE 7.4 Extract from the Results Framework for the Project 'Increasing Climate Change Resilience of Maldives through Adaptation in the Tourism Sector'. (Project Document, ID: 00076855, United Nations Development Programme)

Outcome 2	*Indicator*	*Baseline*	*End of project target*	*Source of verification*
Reduced vulnerability of at least ten tourism operations and ten tourism-associated communities to the adverse effects of climate change	Number of island resorts, tourism operators and tourism-associated communities that report reduced vulnerability to climate risks as a result of guidance provided by the project.	Most tourism operators are concerned about their increased vulnerability to climate change, but do not draw on, or comply with, consistent guidance provided by the government to ensure effective, no-regrets adaptation measures to increase resilience.	By the end of the project, at least ten tourism-associated communities have planned and implemented concrete adaptation projects which reduce the vulnerability of their infrastructure, water, waste, land-use planning or energy management systems to climate-related hazards.	Interviews with community representatives
	Private sector investment in climate change adaptation measures which reduce economic losses in tourism operations and tourism-associated communities from extreme climate events (US$).	Economic losses in tourism-related value chains from climate-induced hazards and extreme events are quantified only after catastrophic events.	By the end of the project, at least ten tourism operators are adopting project guidance to invest in climate-resilient water, wastewater, solid waste and infrastructure management systems.	Qualitative field surveys

7.4 Maladaptation in the tourism sector

If adaptation interventions are not well designed and implemented they can actually increase the detrimental consequences of climate change (Hedger *et al.*, 2008). This is termed 'maladaptation' (see Figure 7.7). Barnett and O'Neill (2010) identify five categories of maladaptation. Business-as-usual tourism development can result in maladaptation, or exacerbate vulnerability to climate change, or both. Careful choice of performance targets and indicators can help avoid maladaptation, including the possibility that a focus on achieving success in the short term may ultimately result in decreased resilience to climate change in the longer term.

Tourism businesses with shorter-term planning horizons may be tempted to focus on adaptive measures that build resilience in the short to medium term, but unintentionally increase vulnerability over time. This might include the diversion of a water supply in order to reduce the risk of immediate shortages. In the longer term, such an action could result in greater water shortages and conflicts between the multiple users of the resource. To manage the risks of maladaptation, businesses need to take a longer-term approach to adaptation.

Examples of maladaptation that could occur in the tourism sector include a resort constructing coastal and flood protection structures that unintentionally disrupt the natural dynamic nature of coastal and river systems, and a hotel installing

FIGURE 7.7 Illustration of the contrast between maladaptation and actions to protect and restore wetlands to facilitate adaptation to climate change. (Aldo Suryadiputra, www.wetlands.org/WatchRead/tabid/56/mod/1570/articleType/ArticleView/articleId/2434/Default.aspx)

cooling or water supply systems that increase energy consumption to unsustainable levels. The OECD (2005) provides more specific examples of possible maladaptation in the tourism sector:

- In Bangladesh, and particularly in the Khulna region, policies to encourage tourism and build tourism infrastructure in vulnerable coastal areas may need to take into account the projected impact of sea-level rise and coastal flooding in order to reduce the risk of maladaptation.
- Plans to increase the resilience of livelihoods by encouraging ecotourism in the fragile Sundarbans mangrove forest, on the delta of the Ganges, Brahmaputra and Meghna rivers, may pose a risk of added stress to a fragile ecosystem that is projected to be critically affected by sea-level rise and salinity.
- Tourism promotion and development policies for Mt Kilimanjaro in Tanzania will need to avoid increasing risks to the ecosystem, given that it is already vulnerable to forest fires as a result of warmer and drier conditions.

7.5 Knowledge sharing and learning

Sharing learning from evaluation is essential in the relatively new and rapidly evolving field of responses to climate change by the tourism sector. The UNWTO has established a website dedicated to climate change, including exchange of information.[9] Interested users can register at a 'Climate and Tourism Information Exchange Service' to download more information. In terms of adaptation, Villanueva (2011) notes that its 'unique nature' calls for experience-based learning from monitoring and evaluating the performance of adaptation interventions. These will provide key insights into adaptive capacity and its links to adaptation processes, and to risk reduction and the enhancement of resilience.

As a result, evaluation plans also need to incorporate communication strategies that identify what the knowledge sharing is intended to achieve, and the target audiences for the activities. Stakeholders in the climate change initiative being evaluated will be a key target audience for an evaluation communications strategy. As Hedger *et al.* (2008) note, the process of being involved in the evaluation itself can facilitate behavioural change since individuals and groups will learn and be motivated by interpreting, understanding and making sense of their own experiences. Importantly, Hedger *et al.* (2008) caution that the communications strategy for sharing evaluation findings should also recognize that evaluation results, especially those for adaptation interventions in the economic and social sectors, are rarely unambiguous, and are likely to be context specific. They also highlight that adaptation knowledge is not a neutral commodity that can be easily applied elsewhere.

The use of information and communications technologies for sharing experiences related to climate change response initiatives is increasing rapidly, though as yet there is no specific focus on tourism. The following are examples of websites that facilitate knowledge sharing and learning processes.

Example	*Description*	*Source*
Adaptation Knowledge Platforms and Networks	Provides links to resources on adaptation practices maintained by the UNFCCC Secretariat and the Nairobi Work Programme Partner Organizations.	http://unfccc.int/adaptation/nairobi_work_programme/knowledge_resources_and_publications/items/5135.php
Climate-Eval	A community of practice dedicated to sharing experiences, improving the quality, and thus increasing the utility of evaluation of climate change and development interventions.	www.esdevaluation.org/gefeo/?q=home
Adaptation Learning Mechanism	Systematically documents and shares knowledge on good adaptation practices and operational guidance.	www.adaptationlearning.net
Global Climate Change Adaptation Network	Mobilizes knowledge and technologies to support adaptation policy setting, planning and practices – see below for regional networks.	www.unep.org/themes/climatechange/Adaptation
AfricaAdapt knowledge sharing platform	A French–English bilingual network facilitating the flow of climate change adaptation knowledge for sustainable livelihoods between researchers, policy makers, civil society organizations and local communities in Africa.	http://africa-adapt.net/AA/
Climate Change Adaptation Knowledge Platform for Asia-Pacific	Supports research and capacity building, policy making and information assimilation, generation, management and sharing.	www.climateadapt.asia
Community Based Adaptation Exchange	Facilitates exchange of up-to-date information on community-based adaptation, including news, events, case studies, tools, policy resources and videos.	http://community.eldis.org/cbax/

Example	*Description*	*Source*
Land, Environment & People Research Centre	Set up as a network for knowledge creation and sharing, the Tourism Theme of LEaP provides a large number of documents and information on climate change mitigation and adaptation in the tourism sector.	www.lincoln.ac.nz/Research-Centres/LEaP/Tourism-Business--Communities/Projects/South-Pacific-Tourism-and-Climate-Change/
weADAPT	Online 'open space' on climate adaptation issues (including the synergies between adaptation and mitigation), allowing practitioners, researchers and policy makers to access credible, high quality information and to share experiences and lessons learned.	www.weadapt.org
Climate and Development Knowledge Network	Supports developing countries to deliver climate compatible development by offering advice and technical assistance, cutting-edge research, strategic knowledge sharing and partnership building.	www.climateplanning.org
Tourism Futures	Climate Change section: bringing together a large amount of information on reports, research, media and events.	www.tourism-futures.org/content/category/19/97/278/
Climate Action Australia	Climate Action Australia has been established to provide the tourism industry with a response to climate change. Climate Action Australia will provide the industry with practical and applied tools for both adaptation and mitigation strategies.	www.sustainabletourismaustralia.com/cert.html
PreventionWeb	Provides information and sharing tools for the disaster risk reduction community to facilitate collaboration.	www.preventionweb.net

Example	*Description*	*Source*
Greenhouse Gas Management Institute	Supports a global community of experts involved in measuring, accounting, auditing and managing greenhouse gas emissions.	http://ghginstitute.org/
The International Ecotourism Society	As part of the Society's advocacy campaign focusing on issues around ecotourism and climate change, TIES and its partners have produced a number of papers, and also disseminate a wide range of case studies.	www.ecotourism.org/site/c.orLQKXPCLmF/b.4835379/k.55C1/TIES_Publications__The_International_Ecotourism_Society.htm

7.6 Conclusion

This chapter has highlighted the importance of knowing whether responses to climate change are effective, and the value of learning such that successful actions are scaled up while actions that are unlikely to achieve the desired results are modified or terminated. In this regard, and at a general level, the understanding, methods and practice of carbon accounting are well advanced, at least relative to adaptation to climate change. A comprehensive assessment of policies dedicated to tourism climate change mitigation is missing, and at this point the resulting learning for tourism has to be based on experiences in other sectors.

With respect to adaptation and carbon accounting in the tourism sector, methods and experience are in their infancy relative to other sectors, though much less so for the latter. Importantly, the tourism sector can learn a great deal from the good practices and lessons learned for adaptation and carbon accounting more generally. The small but growing number of practical examples of evaluating the performance of adaptation and mitigation initiatives by the tourism sector provides a rich resource for learning. However, caution must be exercised when transferring and applying good practice examples and lessons learned as effective responses to climate change are usually highly context dependent, and especially so for adaptation.

8

INTEGRATED RESPONSES TO CLIMATE CHANGE

8.1 Introduction

One of the new key words when talking about climate change is 'cross-cutting'. While climate change itself is a well-understood and clearly defined physical phenomenon, related to changes in the global climate due to changes in the composition of the atmosphere caused by human activity, responding to climate change involves multiple dimensions and literally cuts across many, if not all, areas of social and economic policy, planning and action. The reason why climate change involves everyone and everything is that any human activity contributes to climate change. In turn, changes in the climate impact all aspects of life, both human and otherwise. Prominent examples of the cross-cutting nature of climate change include population growth, human and ecosystem health, the North–South divide, humanitarianism and globalization.

This pervasive nature of climate change often makes it necessary to look at it in an integrated way. Focusing on a particular aspect of climate change, without understanding the broader context, interrelationships and feedbacks, can result in unwanted and potentially counterproductive outcomes. In the context of adaptation to climate change, this is referred to as 'maladaptation' (see Chapter 7). Taking an integrated approach to climate change is challenging, however, as the complexity of both the drivers of, and responses to, the changes increases exponentially.

In the preceding chapters we have chosen to separate the responses of the tourism sector to climate change, treating adaptation and mitigation somewhat independently, albeit while still signalling and illustrating that there are important linkages. This pragmatic approach is by no means unusual as it allows for a more orderly presentation of information. And in reality, policy makers and practitioners are often separated into those with responsibilities for mitigation and for adaptation. But for many tourism operators, the separation between mitigation and adaptation responses is more academic than practical as an integrated approach can often deliver synergistic benefits in terms of both operations and economics.

To illustrate the desirability and benefits of an integrated (i.e. cross-cutting) approach, this chapter presents and discusses four aspects of climate change that relate closely to other important areas of debate and practice (Figure 8.1). First, the relationships between climate change mitigation and adaptation are considered. Trade-offs and synergies will be summarized and case studies provided. The second demonstration of the benefits of an integrated approach concerns the relationship between sustainable development and climate change. Third, opportunities for integrating disaster risk reduction and climate change adaptation in the tourism sector will be explored. Finally, the relationship between energy security and climate change mitigation will be analysed. Examples of how more integrated approaches can deliver benefits for the tourism sector are provided in Figure 8.2.

8.2 Climate change mitigation and adaptation

Introduction

Initial responses to climate change, more generally as well as by the tourism sector, focused on mitigation in the belief that such actions would slow, if not halt, human-induced climate change and thereby the need for adaptation. Human society would simply cope with a limited amount of climate change, just as it had handled natural climate variability in the past. However, in the last decade there has been a realization that plausible mitigation efforts will be inadequate and that, even more worryingly, the climate is already changing in ways, at a rate and with consequences that make adaptation an imperative now rather than something to consider for the future (see Chapter 2).

Numerous studies (e.g. Williams and Geddis, 2010) have identified the benefits of an integrated approach to mitigation and adaptation, but they also recognize that

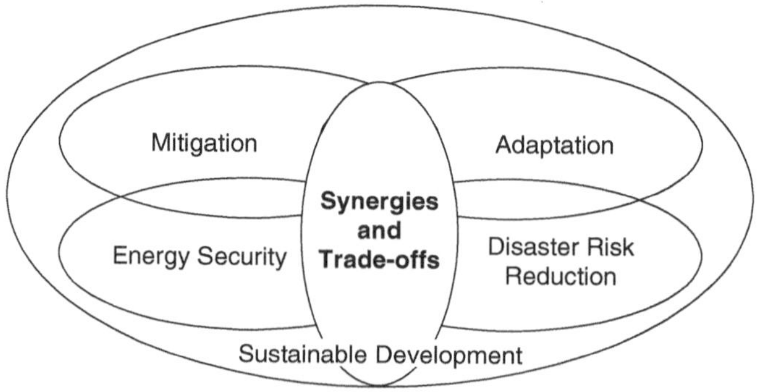

FIGURE 8.1 Synergies and trade-offs arise when tourism responds to climate change in an integrated manner.

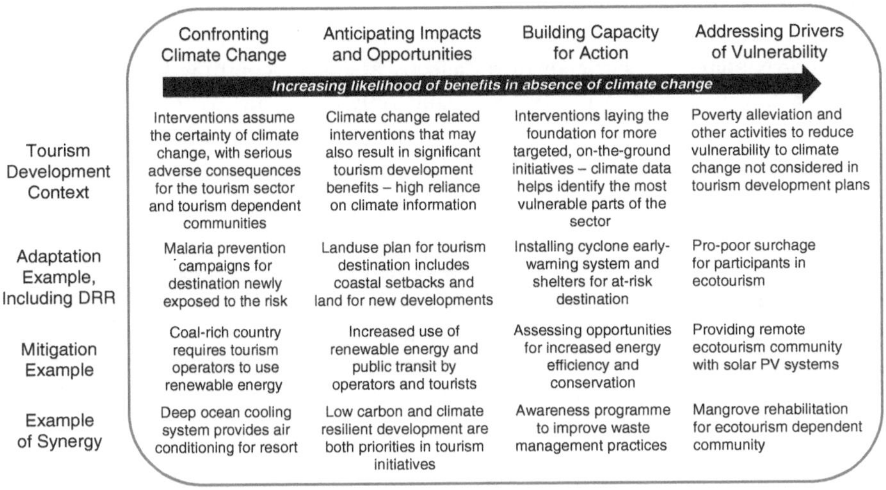

FIGURE 8.2 Responses to climate change, from climate change-focused (left) to development-focused (right), with illustrative examples for adaptation, mitigation and the two combined. (After original framework developed by McGray *et al.*, 2007 and OECD, 2009)

the most suitable approach is context specific. Sometimes it is more appropriate to plan and implement mitigation and adaptation separately, except that it is always important to consider both initiatives in the context of sustainable development (see Section 8.3). The following case studies illustrate that taking a more integrated approach to adaptation and mitigation can be particularly beneficial for tourism-oriented businesses. As a result, there are now numerous tools and other support available to the tourism industry.

Case study: Synergies between adaptation and mitigation for ski resorts

Snowmaking is becoming an increasingly common technique used by ski resorts to supplement natural snow and improve the reliability of their snow cover, thereby extending their season (see Chapter 4). Snowmaking is a relatively expensive process in terms of energy use, and hence large amounts of GHGs are emitted if fossil fuel is used to generate the energy. By using renewable energy the emissions can be reduced, thereby creating a synergy between these adaptation and mitigation initiatives. The National Ski Areas Association of the United States encourages its members to invest in green energy and facilitates a group purchase of green power for its members.

Some ski resorts are now using wind energy to provide 100 per cent of their energy needs, by either using on-site turbines or purchasing power from suppliers who use wind turbines. However, the 1.5 MW wind turbine installed at the Jiminy Peak Mountain Resort in Hancock, Massachusetts, United States (Plate 8.1) provides approximately 35 per cent of the electrical energy consumed annually. The resort was the first private US business, and only ski resort, to invest in its own megawatt class wind turbine. During the winter months, when the wind resource is the strongest but consumption is greatest, the turbine may provide as much as half of the electrical demand. The installed turbine price was approximately US$4 million. The resort received a renewable energy grant from the Massachusetts Technology Collaborative for US$582,000, and it also receives federal tax rebates and renewable energy credits. Payback time is approximately 7–8 years. As of August 2011 the wind turbine had produced 17.6 million kWh of energy since it became operational in August 2007.

The turbine produces 4,600,00 kWh annually, thus offsetting 3,220 tonnes of CO_2, 15 tonnes of SOx (a leading contributor to smog and the chief cause of acid rain) and 4.5 tonnes of NOx (a leading component to smog and a cause of asthma). The Massachusetts Green Communities Act that went into effect in 2009 allows the resort to receive full credit for any exported energy. This means that all energy produced by the turbine is now used on site by the resort.

PLATE 8.1 Wind turbine installed at the Jiminy Peak Mountain Resort. (Vikki Fairbank)

The turbine helps to protect the long-term viability of Jiminy Peak as a resort and therefore reduces the risk of job losses due to changes in power prices. The turbine is designed to shut down during icing conditions, and the machine will not restart until the ice has melted off the tower and blades. Wind turbines have a 30–50 year design life. However, electrical and structural components, including the tower, foundation and interconnection equipment, have a far longer design life. At the end of the useful life of the wind turbine and blades the Resort plans to replace the generator with a new and more efficient unit while using the same tower and electrical components.

The resort is also offsetting its electricity consumption through a series of conservation efforts. Installing new snowmaking technology and upgrading the grooming equipment have reduced energy consumption by 42 per cent and 30 per cent, respectively. Other initiatives include converting ski slope, retail and accommodation lighting to high efficiency bulbs, capturing and recycling heat from compressors, replacing thermostats with secure programmable thermostats for controlling heating and cooling, installing high efficiency motors for heating and cooling and energy conserving kitchen exhaust hoods, and using water cooled towers to reduce conventional air conditioning needs. Collectively these initiatives, along with the wind turbine, have reduced the consumption of grid electricity by 50 per cent (Table 8.1).

The resort is also recycling waste oil from the snowmaking compressors, grooming machines and all other vehicles, amounting to approximately 1,514 litres of waste oil per year. The waste oil is being used to heat the Mountain Operations Building, avoiding disposal problems and further reducing consumption of fossil fuel. Additional energy produced by the turbine will go out into the grid.

Case study: Synergies between adaptation and mitigation for resort hotels

The IPCC has confirmed that the length or number of warm spells, including heatwaves, has increased since the middle of the twentieth century (IPCC, 2011).

TABLE 8.1 Energy Balance Sheet for Jiminy Peak Mountain Resort, Massachusetts, US.

Aspect of energy	*Energy (kWh)*
Baseline consumption	9,404,289
Reduction due to conservation measures	2,342,668
Energy generated by wind turbine and used directly	2,300,000
Energy exported to the grid	2,300,000
Energy consumed from the grid	4,761,621
Net energy consumed from the grid	2,461,621
Net reduction in energy consumed from the grid	73.8%

It has also concluded that it is 'virtually certain' increases in the frequency and magnitude of warm daily temperature extremes will occur through the twenty-first century and it is 'very likely' the length, frequency and/or intensity of warm spells, including heatwaves, will continue to increase over most land areas. These findings have major implications for the hotel industry as it is one of the most energy-intensive sub-sectors of the tourism industry, with about 50 per cent of the overall energy consumption due to air conditioning (EarthCheck, 2010; Bohdanowicz and Martinac, 2002).

Thus the growing use of air conditioning, as a response to increases in high temperatures, including heatwaves, not only increases operating costs but will also increase GHG emissions. This is especially an issue in tropical environments where hotels are already heavily reliant on air conditioning and cooling. Innovative, cost effective and sustainable solutions are therefore required. One such example of innovation is the world's first commercial deep seawater air conditioning system (WWF *et al.*, 2010). It is located at the Intercontinental Resort and Thalasso Spa in Bora Bora, French Polynesia (Plate 8.2). The energy efficient and climate- and environmentally friendly system was designed by Makai Ocean Engineering Inc. of Hawaii. It became fully operational on 1 May 2006.

PLATE 8.2 Intercontinental Resort and Thalasso Spa in Bora Bora, French Polynesia. The pipeline carrying cold ocean water from 900 m depth runs across the reef flat (foreground) to a building located just above the shoreline. As is evident, the installation has no significant visual impact. (R. Bailey, Pacific Beachcomber SC Company, Tahiti, French Polynesia)

Off the reef of Bora Bora, the deepest ocean pipe in the world rises from a depth of 900 m (Figure 8.3). The pipeline is 2 km long and 400 mm in diameter. It brings cold deep seawater at a temperature between 5 and 5.5°C to the surface using capillary pressure – i.e. no energy intensive pumping is involved. By the time it reaches the reef flat the seawater is at a temperature of 6–7°C. An efficient 13 kW pump transfers the cold water on-shore to a non-corrosive titanium heat exchanger connected to the hotel-wide freshwater cooling network that provides air conditioning and refrigeration to hotel villas, restaurants, kitchens, spa, staff residences and other areas. The Resort's Deep Ocean Spa also uses the mineral-rich deep seawater while the desalinated, extraordinarily pure deep seawater, containing mineral salts, is bottled at the Spa. The seawater, which has been warmed to between 12 and 13°C, is returned to the ocean at a depth of 40 m in order to reduce impacts on the ecosystem.

The freshwater network is powered by a 15 kW pump. Together, the two pumps provide cooling that would otherwise consume 300 kW of electricity, thus reducing annual electrical consumption by 90 per cent. This is equivalent to 2.5 million litres of fuel oil per year, and reduces overall energy costs by an estimated 40 per cent, or approximately US$600,000 per year, compared to comparable resorts in French Polynesia. The system also avoids significant environmental impacts, including GHG emissions (by 2,500 tons of CO_2 per year) and the release

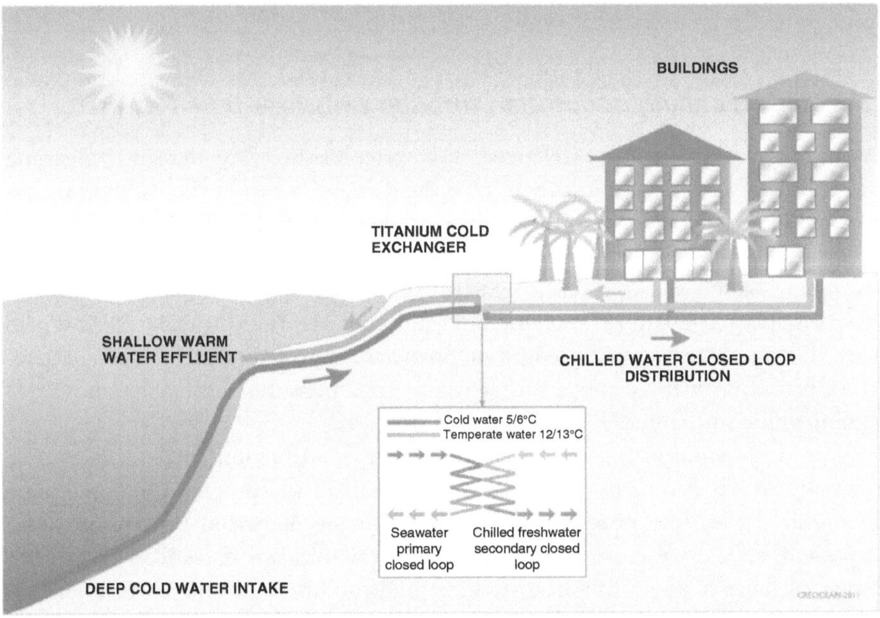

FIGURE 8.3 Diagrammatic representation of the deep ocean water-cooling system at the Intercontinental Resort and Thalasso Spa in Bora Bora, French Polynesia. (Richard Bailey, Pacific Beachcomber SC Company, Tahiti, French Polynesia)

of ozone depleting substances often used in air conditioning systems. At the time the system was being designed energy costs in Bora Bora were US$0.48 per kWh for electricity and US$1.35 per litre for diesel fuel.

The cost of the initial investment was extremely high, at US$7.9 million. This can be compared to US$1.5 million for a conventional air conditioning system for the same hotel. Because it used renewable energy the installation qualified for a 35 per cent tax credit from the Government of French Polynesia. This reduced the payback period to only five years (seven years without the tax credits), allowing for substantial cost savings each year thereafter, despite rising energy costs. Although electricity in French Polynesia is partially subsidized by the French Government, its cost is among the highest in the world, thus contributing to the short payback period.

Prior to the system's installation research was conducted over a period of two years. This highlights that, while deep ocean water-cooling technology has considerable merit, its successful use is dependent on several site-specific factors, including the bathymetry and current regime immediately offshore, the distance the water must be pumped onshore, and the biodiversity. For example, at the sister resort The Brando, pumping the cold water from the ocean surface to the heat exchanger would cost approximately 50 per cent more due to the increased distance involved. Soneva Fushi, a resort in the Maldives managed and owned by Six Senses Resorts and Spas, invested substantially in a deep seawater cooling system, but issues related to the operation and maintenance of the distribution system prevented the pilot programme from continuing (WWF *et al.*, 2010).

Case study: Funding adaptation through emissions-based taxes

Aviation is a sector with a relatively low price-elasticity of demand, meaning that price increases do not greatly reduce the demand for most flights. This makes taxation an unsuitable method for reducing demand, but indicates that it could be suitable for raising revenue. By 2050, the costs of adapting to climate change in developing countries could reach US$100 billion per year, according to estimates from the United Nations Development Programme (UNDP) and the World Bank. Despite international pledges of financial support to developing countries for adaptation as well as mitigation, it is unclear where this substantial amount of funding will come from.

New and additional sources of funding for adaptation are needed. The International Air Passenger Adaptation Levy (IAPAL) is a proposed new purchase tax on air tickets, the proceeds of which would be dedicated to investment in adaptation to climate change. IAPAL would not mitigate the effects of climate change because it does not aim to reduce flight numbers and therefore aviation's contribution to climate change. IAPAL could immediately rise up to US$10 billion annually for adaptation, and considerably more in the longer term (Baker, 2011). The IAPAL follows earlier initiatives, such as the French airline ticket tax, which were set up to raise funds from air passengers for poverty alleviation.

Other examples

Further examples of how integrating climate change mitigation and adaptation relate, amongst others, to the provision of information tools (e.g. guides for operators) or the reduction of GHG emissions of adaptation initiatives are listed below.

Example	*Description*	*Source*
Climate Change Guide, Australia	This guide was designed to supplement a series of free workshops being held across Australia in June and July 2009. The guide has been developed specifically for small to medium-sized operators.	Department of Resources, Energy and Tourism (2009)
Rwenzori Mountains National Park, Uganda	Carbon trading programmes are being tested on the Rwenzori Mountains foothills and could become an alternative income stream for the communities that surround the park. Coupled with adapting and diversifying the tourism product to decrease reliance on high-altitude trekking, these programmes could help to mitigate the damage that climate change has and will continue to have on the economic stability of the communities that surround the park.	Jones (2011)
Implications of Climate Change for Air Transport – A UK Case Study	Flight repositioning represents between approximately 2 and 3.7% of daily emissions of all UK domestic flights; this finding emphasises the importance of future disruptions to national UK emissions estimate.	Pejovic (2008)
Optimising ship operations	A computer program called voyage and vessel optimization system (VVOS) provides the ability to reduce a ship's fuel consumption and CO_2 emissions. Optimization includes wind, waves and ocean conditions.	Ballou *et al.* (2008)
A paradoxical issue in tourist behaviour	High levels of GHG emissions are created by cruise ship tourists visiting the Antarctic – up to eight times higher per capita and per day than average international tourism trips. A survey found no evidence for the hypothesis that the trips develop greater environmental awareness, change attitudes or encourage more sustainable future travel choices.	Eijgelaar *et al.* (2010)

Example	*Description*	*Source*
Links between climate change and electricity generation	Electricity generation is important for powering tourism, but climate change alters disaster risks for electricity generation; this study explores the links between changing disaster risks due to climate change and the impacts on electricity generation.	Urban and Mitchell (2011)
Mitigation, adaptation and resilience – linking business messages on climate change	Investigates ways of linking climate change adaptation with resource efficiency and business continuity advice, by establishing linkages between mitigation and adaptation and resilience measures, and assesses how these considerations can be streamlined into the advice provided by existing business support services.	www.bitc.org.uk/document.rm?id=11937
Climate Change and Tourism Policy in OECD countries	Assessment of both mitigation and adaptation policies implemented in OECD countries.	OECD and UNEP (2011)

Conclusion

The case studies and examples presented above have shown the advantages for tourism if an integrated approach to adaptation and mitigation is taken. However, as was demonstrated, due to the importance of local circumstances care should be taken to ensure the most appropriate combination of approaches is used. Integrated approaches can foster technical innovation and result in considerable cost savings.

8.3 Sustainable development and climate change response

Introduction

Sustainable development had its origins with the Brundtland Commission and concerns the use of natural resources in ways that allows for intra- and intergenerational equity. There are many definitions of sustainable development, with differing emphases on environmental sustainability, social or cultural dimensions or economic development opportunities. As such, sustainability or sustainable development are closely linked to a 'triple bottom line' approach, or even to a 'quadruple bottom line' that explicitly includes climate change as a fourth dimension (UNWTO *et al.*, 2008).

Sustainable tourism is closely related to sustainable development, but is not necessarily the same. Hunter (1998) argued that the sustainability debate in the field of tourism had isolated itself from wider discussions, resulting in the 'emergence of an overly simplistic and inflexible paradigm of sustainable tourism' (p. 850). He redefined the concept as a new paradigm which incorporates a spectrum of approaches to the tourism–environment system, including one where sustainable development is best achieved in the absence of tourism.

Many climate change initiatives contribute to sustainable development (and sustainable tourism), although the boundaries and key drivers are often blurred and sometimes purposefully mislabelled (e.g. for funding reasons). The overlap of multiple motivations is not necessarily a problem, especially in cases where win–win situations emerge. As UNEP and CAST (2008) observe, there is a continuum of initiatives from those that are specifically developed for climate change adaptation to those that address poverty alleviation, capacity building and disaster risk reduction, resulting in tangible co-benefits. Further, the World Resource Institute concludes that 'adaptation needs to be regarded as a "big tent" encompassing the full range of responses ... to managing risk and directly confronting climate change impacts' (WRI in UNEP and CAST, 2008: 26). Similar overlaps and ancillary benefits exist between climate change mitigation and energy security (see Section 8.5), air quality management and health, nature conservation (e.g. deforestation), food security and traffic management.

> Climate change is the pre-eminent geopolitical and economic issue of the 21st century. It rewrites the global equation for development, peace and prosperity. (UN Secretary General Ban Ki-moon)

The remainder of this section will present case studies that focus on sustainable tourism in the context of a beach resort, a global advocacy initiative that integrates sustainable tourism, climate change and peace, and the integrated (and participatory) management of a tourist destination in the face of climate change impacts.

Case study: Climate action as part of sustainable tourism

The Pacific Resort Hotel Group is a boutique resort operator in the Cook Islands that owns and operates three properties, including Pacific Resort Rarotonga. This is a 20-year-old, purpose-built four star resort with 64 accommodation units located on 5.5 acres of beachfront land (Figure 8.4). The resort has bronze accreditation by EC3 Green Globe and is on track to achieve a silver rating.

The vision statement in the hotel's Environmental and Sustainability Policy is:

> To be leaders among Pacific Island accommodation providers by managing our business in a way that demonstrates our commitment to environmental

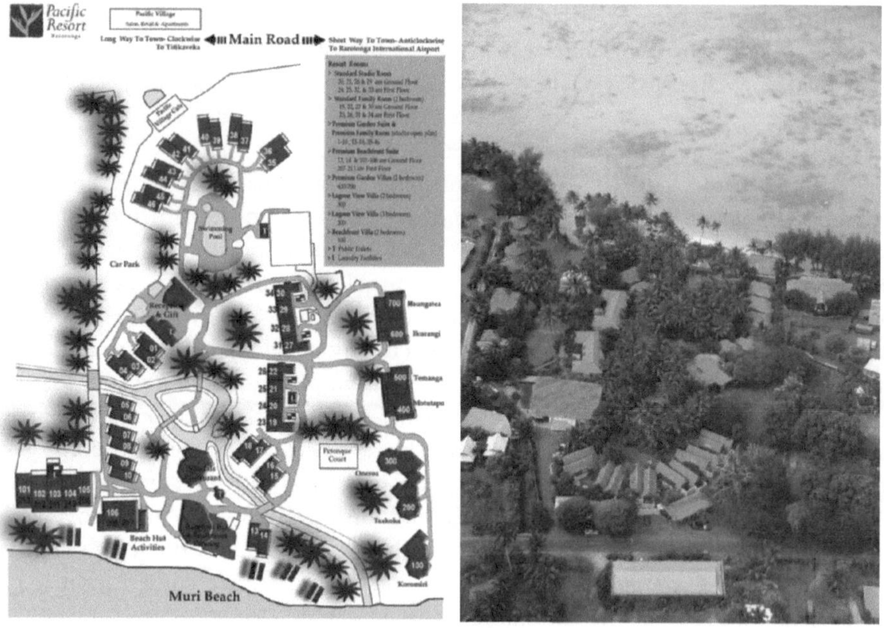

FIGURE 8.4 Pacific Resort Rarotonga, Cook Islands. (Pacific Resort Hotel Group)

> protection, sustainability and stewardship while meeting the expectations of our visitors and the Cook Islands people.

The vision is being achieved in numerous climate friendly and climate smart ways. Guests are familiarized with the Policy, and there are signs highlighting a guest recycling collection area as well as signage in bathrooms asking guests to conserve energy and water. Where practical, the resort collects and treats rainwater, to augment the conventional water supplies.

To the extent possible, locally produced and sourced products are purchased in order to minimize the relatively high GHG emissions that result when items are transported from New Zealand, Australia or other distant countries. To facilitate the use of local produce the resort has established a business relationship with the local growers' association. The intent is to steadily increase the amount of organic/biologically grown produce supplied, minimizing importation and reinforcing the island's 'Go Local' campaign. The resort provides the growers' association with all its green waste, which is used to produce compost for distribution to the growers.

Opportunities to become greener are discussed with suppliers of imported materials. The major supplier of imported food and beverages to the resort collects all cardboard packaging. This is returned to New Zealand for recycling. Otherwise this waste would be disposed of in the local landfill. All waste glass is crushed by a contractor, who uses it for sand.

Diesel generators provide the majority of electricity consumed on the island of Rarotonga, meaning that electricity is expensive (the commercial rate is NZ$0.81 per kWh) and climate unfriendly. As a result, the resort is active in using and researching energy conservation and fuel substitution. All rooms are fitted with key-tag switches which control the operation of air-conditioning and lighting when guests leave the room. All rooms are equipped with solar hot water systems. Standby power generators are maintained regularly to ensure their efficient operation when they are used.

The resort will soon be running its generator on cooking oil one day a week. This will save up to 14 per cent of its electricity costs. It is already running two diesel trucks on biodiesel produced using cooking oil from the resort and two local food outlets. As a result, fossil fuel consumption has been reduced, with savings amounting to NZ$80 per week. The resort is in the first stages of a project to capture methane from its sewage treatment system to use for cooking and for flare gas. A locally based energy specialist has conducted an energy audit for the resort, and is now researching further ways to increase its energy efficiency.

Muri Lagoon, on which the resort is located, is a key attraction for its guests. Reports indicate that water quality in the lagoon has declined in recent years, increasing its vulnerability to the impacts of climate change. The resort was the first tourism operator in the Cook Islands to purchase an advanced sewage treatment plant. Treated grey water is dispersed throughout the resort gardens through a sprinkler system. These initiatives help prevent nutrients seeping into waterways and the lagoon, while also assisting in maintaining the luxuriant hotel gardens. As a result, the resort is a comfortably cool environment during hot weather (with a reduced need for air conditioning). Vegetation mulching is creating deeper soil profiles. These provide a buffer in times of drought and also help prevent nutrients reaching the ground water and leaching into the lagoon.

The resort is a sponsor of 'Lagoon Day', a local educational event designed to raise awareness of the ways in which lagoon health can be impacted. As part of this event the resort runs tours of its recycling, water treatment and sanitation systems. The Pacific Resort Hotel Group has also financially supported Cook Islands youth representatives to attend the Conference of the Parties to the UNFCCC. These initiatives demonstrate that not only is the Pacific Resort very active in environmental management (including both climate change mitigation and adaptation) but it also sees its role as raising awareness, contributing to community initiatives and 'making a difference' towards the sustainable development of the Cook Islands.

Case study: Global advocacy initiative for climate action and world peace

The 'Lusaka Declaration on Sustainable Tourism Development, Climate Change and Peace' (Figure 8.5) is a useful example of a call for action across climate change mitigation and adaptation as part of an integrated approach to tourism, peace and sustainability. The actions identified in the declaration involve the tourism industry, governments and travellers. Initiatives include reducing GHG emissions, reducing pressures on natural resources (e.g. water), and preparing for natural disasters. One government recommendation suggests incentivizing 'tourism projects that utilize green designs/technologies and contribute to creating sustainable livelihoods, poverty reduction and the UN Millennium Development Goals', and only approving 'incorporate construction designed to withstand severe storms and other effects of climate change'. The proposed actions are also intended to appeal to people's values and ethics. The Declaration builds explicitly on the 2007 Davos Declaration by the UNWTO.

The Lusaka Declaration stems from the 'Meeting the Challenges of Climate Change to Tourism in Africa and the Developing World' conference organized by the International Institute of Peace through Tourism, and held in Lusaka, Zambia, in May 2011. This was the first global conference on climate change and tourism that took a holistic approach to tourism and climate change, and that purposefully focused on developing countries. It became clear at the conference that avoidance of unsustainable development (e.g. building styles that are not appropriate for a given location) is the first critical step towards successful climate change responses.

And therefore call on all sectors of the travel and tourism industry to:

1. *Support the leadership role of UNWTO, UNEP and WMO in the specific actions called for in the Davos Declaration: Climate Change and Tourism – Responding to Global Challenges.*
2. *Give emphasis to energy conservation and aggressively make the transition to renewable and clean energy sources as a foundation for carbon/GHG neutral development strategies and a 'Green Growth' transformation.*
3. *Continue to implement socially and environmentally responsible tourism practices that benefit local destinations and enhance both cultural and natural heritage, giving particular emphasis to fair trade in tourism practices, tourism that contributes to poverty reduction and tourism that contributes to the restoration of ecological systems.*
4. *Provide for quality tourism experiences that include meaningful interactions between visitors (guests) and the host community and a greater understanding and appreciation of local cultures, customs and traditions.*
5. *Implement environmentally responsible development and operational practices, particularly in vulnerable areas, including coastlines, low-lying areas and mountain regions.*
6. *Develop and implement waste management plans that give emphasis to reducing, recycling and re-using.*
7. *Incorporate quadruple bottom line reporting with climate change alongside environmental stewardship, social responsibility and economic value.*
8. *Actively engage local communities in planning and decision-making, empower women and youth and embrace the wisdom, knowledge and values of indigenous peoples.*
9. *For the transport sector, and particularly the airline industry, continue efforts to improve engine efficiencies and ultimately develop engines that operate with clean and renewable energy sources.*
10. *Recognize the important role of media, educational and religious institutions as agents of change.*

FIGURE 8.5 Excerpt from the Lusaka Declaration on Sustainable Tourism Development, Climate Change and Peace.

Further, a wide range of case studies have highlighted that 'carbon management' is rapidly becoming part of good business practice, rather than merely a moral response to political pressures on reducing GHG emissions.

The importance of institutional arrangements, even those including the media, became apparent for both mitigation and adaptation measures. Partnerships and communication between key stakeholders from all levels are essential, for example between governments and tribal or indigenous communities. Finally, an insight from the Lusaka conference (and also anchored in its declaration) was that tourism development could be interpreted as a climate change adaptation measure, especially in situations where tourism is relatively less vulnerable to climatic changes than other sectors (e.g. agriculture).

Case study: Sustainable management of a tourist destination

The Island of Oléron is the most southern of the islands off the French Atlantic coast. It has been an important tourism and holiday destination since its connection to the mainland by a bridge in 1966. Beaches, surfing, horse-riding, forests and the many heritage buildings are the main attractions. As is the case for many coastal areas, the island faces numerous challenges to its sustainability: fragile environmental, social and economic systems are threatened by a number of pressures including those related to: climate change; seasonal mass tourism; increasing conflicts between the different uses of the coastal environment; inflation of the price of land; a decrease in the number of working people in the population; and a loss of local identity (European Commission, 2010).

Duvat (2011) describes the processes involved in developing and implementing the Oléron Coastal Quality Action Plan, as well as the findings of studies carried out to gather public opinion on the resulting changes in coastal quality. Coastal quality includes both environmental quality (e.g. condition of ecosystems and resources, quality of landscape and the living environment) and the quality of the tourist experience (e.g. heritage value, amenities provided and information delivered). The case study therefore provides a useful illustration of how managing a tourism asset links to wider development issues and processes. It also shows, in a practical way, the linkages between many of the topics covered earlier in this book, such as policy making, planning, implementation, monitoring and evaluation.

An integrated coastal zone management (ICZM) policy was adopted by the Marennes Oléron District in the early 2000s. The first action in the ICZM project was a participative initiative that resulted in the adoption of the Sustainable Development Charter. This outlined the strategy for the territory for the next 15 years. It led in 2003 to the designing of a 'Beach Scheme' to improve both the condition of coastal ecosystems and the quality of the tourist experience.

Subsequently, the District's ICZM project (2006–2008) provided a coherent approach, including rallying diverse stakeholders around a territory project as well as providing them with common principles and tools. The project resulted in the adoption of key planning documents, including the Oléron Coastal Quality Action Plan. This was an outcome of active stakeholder participation. First, the project carried out a survey to identify the main concerns of stakeholders and their opinions on how these might best be addressed. Second, the project developed and used tools for knowledge sharing (e.g. monitoring chart of issues and early indicators of sustainable development) and territorial development evaluation (e.g. beach quality assessment). The Oléron Coastal Quality Action Plan was adopted as the first action plan to implement the territorial project through an ICZM approach. Its main objective is to strengthen the preservation of coastal ecosystems without obstructing economic development, including tourism. The Plan, which was launched in late 2007, directly benefited from the earlier ICZM initiatives.

In the design phase of the District's territory project the lack of scientific knowledge and assistance to stakeholders from the scientific community were identified as major shortcomings for the implementation of public policies. This problem was addressed through the establishment of a funded research programme designed to assist stakeholders in the implementation of their beach development project. The research programme runs from 2008 to 2012. It first aims to clarify the scope of coastal quality, identify its main drivers and establish the methods to assess it. The second component involves coordinated production of data and an annual work programme to address the specific and changing needs of stakeholders. The collective building of knowledge and continuous sharing of scientific results facilitate development of a common rational vision of coastal quality and how quality targets can be achieved.

The Plan includes limited redevelopment of La Giraudière beach on the southwest coast of Oléron Island. The beach is protected by various regulatory standards due to its high ecology value, fauna and flora, and is classified in the 'leisure' category (Table 8.2), which allows only limited development to take place. While the beach is very popular (3,724 visitors in summer peak hours), rapid erosion is resulting in the loss of 20 m of dunes on average each year. This is due, in part, to intense marine and wind erosion. Dune retreat is reducing the beach's accessibility, with the decreasing area of foreshore forcing users to retreat to the dunes at high tide, thereby accelerating their destruction. The changes could jeopardize the popular practice of land sailing once the rocky substrate replaces the beach. Erosion also threatens to destroy the parking area.

In this context, the municipality applied for the problem to be remedied. The plan includes relocation of the car park into the wooded dune to protect it from erosion, with a simultaneous development of public transport to maintain beach accessibility, as well as the consolidation of facilities (e.g. clubhouse, toilets) onto a dedicated platform. However, implementation has been delayed. In fact, the implementation of five of the six beach redevelopment plans has been delayed. Reasons are mainly of an administrative, regulatory and financial nature, with all of these being linked. Importantly, the overlap of administrative and management

TABLE 8.2 Summary of strategic directions for the beach destination scheme of the Oléron Island Beach Plan. (After Duvat, 2011)

Beach type	*Access*	*Planning objectives*	*Nature of beach facilities*															*Cleanliness of beaches and dunes*	*Protection of beaches and dunes*
			General attributes												*Special attributes*				
			Reception							*Security*		*Environmental protection*			*Wheel chair access*	*Naturism*	*Dogs*		
			(1)	(2)	(3)	(4)	(5)	(6)	(7)	(8)	(9)	(10)	(11)	(12)					
Nature	Limited	Few or no facilities	Yes	Yes	No	No	No	No	No	Yes	No	Yes	Yes	Yes	Yes	Three beaches are authorised	No	Cleaning should be considered at the island scale and adapted to the nature of beaches	Protection from erosion should be implemented on a case by case basis
Environment discovery	Controlled	Few facilities	Yes	Yes	Yes	No	Yes	No	No	Yes	No	Yes	Yes	Yes	Yes		No		
Leisure	Controlled	Limited facilities	Yes	Yes	Yes	Yes	Yes	Yes	Yes	Yes	Yes	Yes	Yes	Yes	Yes		No		
Resort	Open	Facilities integrated into the urban fabric	Yes	Yes	Yes	Yes	No	Yes	Yes	Yes	No	Yes	No	Yes	Yes		No		

(1) Signage; (2) Parking; (3) Toilets; (4) Showers; (5) Picnic; (6) Snack outlet; (7) Emergency station; (8) Emergency access; (9) Orientation symbols; (10) Beach markings; (11) Access markings; (12) Awareness raising signage

boundaries and regulations causes the extreme fragmentation of the territory, making it difficult to implement a project in a standard way at a more local scale. At the same time as public expenditure is being reduced drastically, the strengthening of environmental remediation standards and the technical requirements of the Charter are contributing to the increasing cost of development initiatives.

Some beach plans have been suspended for two years due to the failure of stakeholders to raise the required funding. Financial problems are being resolved in two ways, through the resizing or rescheduling of beach plans, to make them compatible with the available financial resources, and/or through increased fund-raising efforts. Budget constraints have also emphasized the need to share the financial burden between partners, reinforcing the importance of a relationship building process that brings all stakeholders together.

Beach plans blocked by conflicts of interest have required long negotiations between local authorities and government services, with all involved parties having to make considerable compromises. Solutions included the abandonment of some parts of the project, or providing compensation. Learning has resulted in increasing consultation in the early stage of projects, contributing to a better integration of government services into the decision-making process. The main benefit of the resolution of conflicts has been the strengthening of the management capacity of stakeholders. The resolution of conflicts has also allowed the realization of major projects to improve environmental quality.

Recently, studies have been carried out to gather public opinion on the results of implementing the Oléron Coastal Quality Action Plan, such as for the beach of La Brée-les-Bains (Plate 8.3). The beach is located on the northeast side of Oléron Island and is relatively sheltered from the swells and prevailing winds. The beach was the first to undergo a comprehensive redevelopment, involving raising the standards of all facilities (e.g. first aid station, toilets, bins) to the levels required by the Charter.

A baseline opinion survey had been conducted, involving interviews with 418 beach users as well as with professionals (lifeguards, instructors of activity clubs, snack outlet employees). A follow-up evaluative survey of 200 beach users and other stakeholders was conducted in mid 2009. The survey was designed to determine their views on the completed developments. It also aimed to determine whether or not the beach users were favourably disposed to the strengthening of the urban nature of the beach through the development of a promenade and of seaside entertainment, as part of diversifying the range of available beach types (see Table 8.2). Users had high expectations regarding beach quality. Nearly 90 per cent of beach users were satisfied with the redevelopment carried out to meet the Charter's requirements, but showed disappointment with the closure of the children's club in 2009, and strongly criticized the municipality for not removing algae accumulations on the foreshore. Opinions were sharply divided regarding the plans to strengthen the urban aspect of the beach: 60 per cent of visitors would like more entertainment on site, while 50 per cent supported pedestrianization of

PLATE 8.3 Beach of La Brée-les-Bains, Île d'Oléron, France. Note how the erosion control works limit the useable area of beach. (V. Duvat-Magnan)

the beach edge. However, 36 per cent and 42 per cent, respectively, were opposed to these ideas.

Other examples

Sustainable tourism, both academically and practically, has a long tradition, and there are many examples of projects and initiatives that seek to achieve a 'more' sustainable (tourism) development. Most of these projects have a direct or indirect link to climate change initiatives. For example, sustainable tourism projects that focus on resource efficiency are likely to contribute to reducing GHG emissions. Similarly, community-based projects are likely to increase the resilience of local stakeholders and groups. As such, they will benefit through adaptation to additional challenges such as climate change. Some sustainable (or eco-) tourism projects include adaptation measures directly, for example those relating to water consumption and quality, nature conservation (e.g. coral reefs), reforestation or design of new infrastructure and buildings. In some cases tourism is used as an income generator to avoid less sustainable practices, and to improve overall sustainability of destinations, communities and ecosystems. A small number of examples are shown overleaf.

Example	*Description*	*Source*
Green Hotelier publication	Online magazine by the International Tourism Partnership to inform hoteliers about environmental initiatives and to share lessons learned.	www.greenhotelier.org/
Shimla Declaration on Sustainable Himalayan Development	Initiative in the Indian Himalayas that integrates climate change and sustainable development with specific reference to eco-friendly tourism and pilgrimage, green transportation, management of water resources, and energy security, amongst others.	http://himachal.nic.in/environment/pdf/HimalayanCMConclaveDecl.pdf
Blue Yonder, India	Tourism operator that integrates community development, responsible tourism and climate change responses (e.g. local tree planting, nature conservation, education).	www.theblueyonder.com/policies.htm
Greening Mexican Tourism	MARTI is a collaboration of six organizations working to create more sustainable practices across the tourism industry in the Riviera Maya, Cancun and Cozumel. By engaging leaders from the government, nongovernmental organizations, hotels, tour operators and cruise lines, MARTI provides a forum for these stakeholders to work across the tourism supply chain and integrate their efforts.	www.conservation.org/fmg/articles/pages/greening_tourism.aspx
Caribbean Tourism Organization	Presentation that seeks to integrate climate change challenges and opportunities for local communities: 'Climate Change and Community-Based Tourism – Key Policy Recommendations for Sustainable Tourism'.	www.onecaribbean.org/content/files/CTO%20GailHenryColombiaClimateChangePrestn.pdf

Example	*Description*	*Source*
Ibo Island Lodge, Mozambique	With the African tour operator Wildlife Adventures, Conservation International established the ecolodge in 2006. Ibo is home to some of the most pristine and unexplored island reef ecosystems in the Indian Ocean and provides refuge for ten globally threatened bird species and five globally threatened turtle species. But poverty among the island's 4,000 residents was driving slash-and-burn agriculture and habitat loss, unsustainable fishing, and illegal hunting of marine turtles and sharks. The lodge provides income and (indirectly) reduces ecosystem vulnerability to climate change.	www.conservation.org/learn/culture/ecotourism/destinations/Pages/mozambique_ibo_island_lodge.aspx
Tourism and food	Tourism is a major consumer of agricultural products and manufactured food. As such it is has an important role in reducing GHG emissions of food related 'supply chains'.	Gössling *et al.* (2010b)

8.4 Disaster risk reduction and adaptation

Introduction

There is strong evidence (e.g. IPCC, 2007a, 2011; McMullen and Jabbour, 2009; Richardson *et al.*, 2011; Webster *et al.*, 2005) of an increase in both the observed frequency and intensity of weather- and climate-related hazards, such as heavy rainfall, that frequently result in flooding; droughts; high sea levels often exacerbated by storm surges; and possibly tropical cyclones. These increasing trends, which are consistent with those anticipated as a result of global warming, have important implications for disaster risks. Weather-related disaster risks are increasing, with more devastating impacts on communities than expected.

Moreover, there is also growing evidence that, at least in the short to medium term, many important impacts of climate change may manifest themselves through a change in the frequency, intensity or duration of extreme events (IPCC, 2011). Thus, while responses to climate change may initially have been framed by a longer-term outlook, there is now at least as much emphasis on the present, as well as the immediate future. Similarly, the disaster risk management community is moving rapidly from looking only at historic and current risk, to considering future risks. Responding to the future changes in extreme events will require bolstering disaster risk management as a first line of defence.

Since the consequences of climate change include increasing disaster risk, there will be an increasing burden on development programmes and initiatives, including those related to tourism. This represents important progress in understanding since, until recently, it has been difficult for disaster risk management to disassociate itself from a focus on historic disaster events, while for several decades climate change was seen as an environmental rather than a development issue.

Implications for tourism

While tourism development planning and processes should anticipate climate extremes, variability and change to the greatest extent possible, there will always be weather and climate 'surprises' – unforeseen events including climate-related disasters, such as tropical cyclones (also called typhoons or hurricanes). Planning for 'climate surprises' is becoming increasingly important, and is yet another example of the benefits of integrating adaptation and disaster risk management in tourism policies and plans.

Promotion of a comprehensive disaster risk reduction approach that includes preparedness must therefore be a key component of climate change adaptation in the tourism sector, with adaptation processes drawing on approaches to disaster risk reduction, as well as tackling gradual changes and new hazards. This means that climate resilience must be increased in both the short and longer terms, by focusing on the immediate climate and disaster risk reduction needs, while also building capacity for dealing with longer-term adaptation challenges. Importantly, the basis for adapting to the future climate lies in improving the ability to cope with existing climate variations and extremes. Climate change projections inform this process, to ensure that current coping strategies are as consistent as possible with the anticipated changes in the future climate.

Tourism often occurs in remote natural areas and coastal environments, both of which are especially exposed to natural disasters or other crises, and where there is reduced capacity to plan, prepare and respond. Tourism is also highly dependent on the functioning of critical infrastructure such as transportation networks, electricity, sewage, telecommunications and water (Ritchie, 2004; Orchiston, 2010). Tourism in remote areas is also dependent on supply chains, for example for the provision of food. Natural disasters, such as those that are a result of extreme weather events, can have major effects on tourist safety and arrivals in the long term. A disaster can lead to a tourism crisis, a situation that requires exceptional response measures by the affected community, and which challenges the operation and survival of an organization or destination. Several countries have invested resources into developing tourism-specific disaster and crisis plans. Australia's National Tourism Incident

Response Plan (Commonwealth of Australia, 2007), for example, 'sets out a process for delivering a coherent national response aimed at minimizing the impact of incidents on the Australian tourism industry' (p. 7). Based on this Australian national plan, Tourism Queensland (2007) developed a template for regional tourism organizations to design their own crisis plans (see Chapter 3). This template suggests that a response and recovery plan is structured into actions that relate to preparedness and response, and those dealing with recovery. Marketing and business assistance are key elements of recovery.

Examples of recent weather disasters and tourism crises include the Pakistan flooding, the Queensland flooding (World Bank and the Queensland Reconstruction Authority, 2011), the Russian heatwave and fires, and the European and North American snow storms. Research on the 2006–2007 bushfires in Victoria, Australia, revealed that the disaster had a significant, long-term impact on tourism. While the operators themselves felt that they were well prepared and informed, the media reporting was considered to be too negative and to have unnecessarily increased the detrimental impact on visitor arrivals (Sanders *et al.*, 2008) (Figure 8.6). The existence of prolonged media attention was identified as a key challenge to recovery, alongside a lack of explicit forward planning by tourism businesses.

B World

THE PRESS, Christchurch Tuesday, October 25, 2005

Race to rescue tourists

Mexico is scrambling to rescue thousands of tourists trapped in stifling shelters in its hurricane-hit Caribbean beach resorts before food, water and medicine shortages cause chaos.

Hurricane Wilma caused massive damage in Cancun and at other resorts along the "Maya Riviera" with three days of wild winds and rain. Hotels and streets were flooded and littered with debris, and homes and stores were destroyed.

President Vicente Fox said yesterday that the airport in Cancun would not be open for at least two more days and his government would send buses to pull out stranded tourists, starting today.

Seven people were killed in the storm, many fewer than feared given Wilma's size and ferocity. Mexican authorities helped limit the human cost by ordering tourists out of beachfront hotels and putting them in makeshift shelters.

"There is huge devastation. This hurricane has provoked a tremendous impact. But Mexico has experience and it was demonstrated right from the beginning, saving lives," Fox said.

It was not clear how many tourists were stranded but one senior police official estimated there were about 20,000 just in Cancun.

Many were short of food and water and becoming increasingly frustrated yesterday as they faced a fourth night in cramped shelters with no electricity or running water.

Most of their luggage was still in hotel rooms they were unable to get back to, and most had given up hope of relaxing on white sand beaches or scuba diving in coral reefs.

No shopping: Cancun's shopping centre lies gutted after Hurricane Wilma's visit.

Photo: Reuters

MEXICO

"People are starting to get sick. Some of the elderly people are becoming ill," said Doug Ruby, a United States tourist with his wife. "We've got to get out."

Another kind of chaos took over yesterday as police shot into the air to scare looters away from a shopping centre, and looters responded by throwing rocks and chunks of concrete.

Military officials and police stood guard outside many businesses and set up checkpoints to seize looted goods. Still, some people, hungry and unable to find anything open, began taking things they needed. One group pushed carts against the boarded up windows of a grocery store in an attempt to break in.

At a convenience store, Cancun resident Alex Aguilar took batteries and aspirin. "The window was broken, so we just went in and got what we wanted," he said.

Dozens were arrested.

"It's chaos," said fire official Gregorio Vergara. "They are taking things all over the city."

Others waited in long lines at the few stores that were open. At one only accepting pesos, tourists without local currency offered $US100 bills for $US5 calling cards.

Downtown, the city handed out food packages that included rice, beans, crackers and cooking oil, and people stood in line for blocks to collect.

The United States (US) Embassy was sending consular officials to shelters to help tourists prepare to leave. The US Government also offered to donate $US200,000.

The storm was a devastating blow to Mexico's $US11 billion foreign tourism industry.

—Reuters-AP

FIGURE 8.6 Media coverage of cyclone Wilma (2005) in the Christchurch *Press*, New Zealand. The coverage focuses on negative aspects and safety concerns for tourists.

The appointment of a recovery officer was suggested as a measure to address these issues. Also, the allocation of contingency funding in case of future events (by a Regional Tourism Organization, for example) was seen as a useful measure to aid recovery. Funding could be used, for example, to publicize the messages of satisfied post-bushfire visitors.

Case study: Implementing integrated disaster risk reduction and climate change adaptation at a tourist destination

A very informative case study in UNISDR (2010) describes an integrated approach to disaster risk reduction and climate change adaptation that has been undertaken by Hermanus, South Africa, an important tourist destination. This seaside town, with majestic mountaintops as a backdrop, is a popular holiday destination close to Cape Town Plate 8.4. The natural beauty of the area is an outstanding asset, and includes South Africa's first biosphere reserve as well as arguably the best land-based whale watching in the world. The area is most famous for being one of the destinations for the southern right whales that migrate along the coast of South Africa, and where they calve and nurture their young. Apart from whale watching, the area offers numerous activities such as hiking, golf, fishing, boating and horse-back riding. There are three Blue Flag beaches.

Thus tourism is a major economic driver in the area, with a fourfold increase of the population during the holiday season. This influx places a great strain on the existing municipal services and roads infrastructure. For example, over the years, water use in Hermanus has become unsustainable. Due to rapid and seasonal population growth in this tourist area, water demand in Hermanus increased from just over 0.5 million m^3 per year in 1960 to five million m^3 in 2008. The total sustainable yield for the water distribution is seven million m^3. It is estimated that, based on 4 per cent and 6 per cent annual growth rates, water demand will exceed the sustainable yield by 2017 and 2014, respectively (Overstrand Municipality, 2010). This unsustainable situation has been exacerbated by a decline in rainfall since 1997. It is anticipated that there will be further declines in rainfall in the future, as a result of global warming (Engelbrecht *et al.*, 2009). Prompted by the consequences of a drought in 1966–1977, the Overstrand Municipality has escalated work to reduce water use and to find new sustainable water sources.

The Hermanus Water Resource Development and Management Programme was launched in 2001. Its goal is to avoid water shortages and reduce drought risk. The Hermanus programme has adopted the principles and conceptual approach that underpin the policy and legislative platform built by the South African National Department of Water Affairs and Forestry. This is in turn aligned with the Hyogo Framework for Action, a ten-year plan to make the world safer from natural hazards. The Framework was adopted by 168 Member States at the World Disaster Reduction Conference, held at the United Nations in 2005.

The Hermanus Water Resource Development and Management Programme is coordinated, funded and implemented by Overstrand Municipality. In partnership

PLATE 8.4 View from tourist accommodation at Hermanus. (Villa Blu)

with many stakeholders, the Programme is implementing various initiatives such as commissioning new water sources, reducing water wastage and clearing invasive alien plants. Hermanus has actively reduced water demand through the introduction of a stepped tariff structure, a community awareness campaign, a water-wise garden project and accelerated infrastructure maintenance to reduce wastage from leaks. These initiatives are not only resulting in an easing of demands on the water supply system, but are also resulting in significant community engagement, job creation and community development.

More water sources have been identified by accessing groundwater in a confined, fractured-rock aquifer. Prior to this project, the Municipality had no experience of extracting large volumes of groundwater. Therefore, there was some uncertainty, and even scepticism, about the project. The uncertainty centred on the long-term sustainability of the project and the impact it could have on the environment and other groundwater users in the area. This challenge was overcome by establishing the Groundwater Monitoring Committee, consisting of representatives from all the stakeholders. This had the effect that stakeholders felt that they were included and part of the project. Another strategy was to move slowly in the beginning and to gather baseline information.

Treated effluent from the Waste Water Treatment Works is now re-used for irrigating the golf course and sport fields. Previously the treated effluent was discharged into the sea, and potable water was used for irrigation. This initiative has created a new source of revenue for the Municipality. During the most extreme periods of water scarcity, new property development was restricted, including postponement until additional, sustainable water sources were identified.

The overall approach adopted by Overstrand Municipality is transparent, inclusive, sustainable and provides a long-term solution to the problem. It has built on a sound policy and legislative platform, has demonstrated the advantages of a long-term view for building informed community involvement, and has encouraged innovation and the application of the best scientific expertise.

Other examples

Examples of how risk reduction is approached in various destinations are compiled below.

Initiative	*Description*	*Source*
Bushfires disrupting tourism in national parks, Australia	The ways in which tourism operators have reacted to major bushfires in the Alpine National Parks of Victoria have provided the basis for planning for crises and disasters, particularly in terms of understanding the impacts that a crisis or disaster can have on tourism marketing and in relation to planning and managing marketing communication activities after a crisis or disaster.	Sanders *et al.* (2008)
Northland, New Zealand	Destination Northland is developing a Visitor Incident Plan that complements and enhances the activities and plans by the New Zealand Civil Defence and Emergency Management Services.	Becken and Hughey (submitted)
Bahamas	The Bahamas Ministry of Tourism 'coordinates activities to ensure that the tourism industry is equipped to effectively respond to and recover from any disaster impacting the nation' (p. 37). An example is the Hurricane Preparedness and Response Plan, and the establishment of a Tourism Emergency Coordinating Committee.	UNEP and CAST (2008)
Asia Pacific Economic Cooperation, Risk Management Guide	APEC have joined together to address disaster risks for tourism, and developed authoritative guidelines for managing crises and disasters based on the 4Rs (reduction, readiness, response, recovery).	AICST (2006)
Disaster Handbook	Compiled by the World Tourism Organisation to provide practical advice and templates to tourist destinations, businesses and visitors on what to do before, during and after a natural disaster.	WTO (1998)

Initiative	*Description*	*Source*
Crisis Essentials	Practical advice (including checklists) for tourism businesses, provided by Tourism Victoria, Australia.	Tourism Victoria (2010)
Caribbean Regional Framework for Achieving Development Resilient to Climate Change	Seeks to strengthen climate resilience in sectors, including tourism, by supporting the dissemination of successful adaptation and disaster risk reduction measures, some of which are already being implemented in the region.	Caribbean Community Climate Change Centre (2009)

Conclusion

Over the last decade there have been several instances where extreme events have severely affected tourist destinations. In response, there has been an increased interest in assisting tourism operators and associated communities in preparing for disasters and integrating tourism into disaster management frameworks (e.g. UNEP and CAST, 2008). While disasters are caused by a wide range of extreme events, those related to hydro-meteorological and climatological hazards are likely to be exacerbated in the future due to the influence of global warming. New approaches to disaster management and risk reduction are required. Increasingly, civil society and other organizations (e.g. trade unions or economic sectors) are included in disaster management activities. Hence, actions by players that are not formally part of civil defence are no longer considered as disruptive, but are embraced as important additions to the risk management network. This is a great opportunity for tourism to be formally and informally involved in disaster management.

However, despite the vulnerability of tourism to natural disasters and their relevance in forming part of disaster management, it appears that the tourism industry tends to be poorly prepared for disasters (Cioccio and Michael, 2007). Crisis preparation or disaster risk reduction do not typically form part of tourism business practices, even though several attempts have been made to formalize disaster frameworks for tourism (e.g. Faulkner, 2001). This section has presented a case study and several examples which highlight the progress tourism has made to link climate change adaptation and disaster risk reduction.

8.5 'Peak oil' and climate change

Increasingly, peak oil and climate change are being considered together, rather than separately (Becken, 2008a; Verbruggen and Al Marchohi, 2010). As Hughes and

Rudolph (2011) note, 'Accommodating the increasing future demand for secure supplies of liquid fuels, while minimizing its impact on the climate, will require energy policies to be developed that address both energy security and climate change, thereby achieving mutually reinforcing benefits.' The following section briefly outlines the current discussion around a peaking of global oil production, its implications for tourism, and the relationship between energy and climate policies.

> The next five years will see us face another crunch – the oil crunch. This time, we do have the chance to prepare. The challenge is to use that time well. (Sir Richard Branson, Founder of the Virgin Group)

Background on peak oil

Tourism is very oil intensive, depending heavily on an affordable supply of liquid fuel. The aviation industry alone consumes 243 million tonnes of fuel per year, amounting to 6.3 per cent of world refinery production (Nygren *et al.*, 2009). Oil prices increased sharply in 2007 and 2008 (with an all-time high of US$147 per barrel of crude oil). After a drop following the 2009 global recession, they increased once again in 2010 and 2011. While a large number of factors influence the price of oil in the short term, it is argued that prices are ultimately driven by limited resources. Estimates of so-called 'Ultimate Recoverable Resources', and the resulting paths of global oil production, differ (UK Energy Research Institute, 2009). But there seems to be a consensus that conventional oil (i.e. oil that is easily accessible with current technology) is likely to decline in the near future (Greene *et al.*, 2006; Hirsch, 2008; Sorrell *et al.*, 2010). Estimates made by the International Energy Agency (IEA) and the US Energy Information Administration (EIA) and others, that production of conventional oil will continue until at least 2030 have been found to be 'rather extreme and implausible' (Jakobsson *et al.*, 2009).

In addition to conventional oil resources, there are so-called 'unconventional sources'. These feed into a variety of conversion processes that produce a liquid fuel similar to conventional oil. Examples that are often discussed include Canada's tar sands, Venezuela's heavy oils and a coal-to-liquid conversion processes whereby coal is converted into a liquid fuel. Alternative fuels can also be made from biological sources, for example sugar cane or corn. Despite the large number of drawbacks of biofuels, the IEA is projecting that biofuels will meet between 3 and 10 per cent of the world's liquid biofuels demand by 2035 (Hughes and Rudolph, 2011). The problem with all unconventional oils is that the cost of production is substantially higher, and the energy return for the investment is much lower.

This means that per-barrel prices will increase dramatically, as will the embodied energy and hence carbon dioxide emissions. Accordingly, some analysts argue that the production of oil is only a matter of investment and price, and it is the carrying capacity of the atmosphere (in terms of carbon concentrations) that is the real bottleneck, not the exhaustion of oil reserves and resources (Verbruggen and Al Marchohi, 2010). This has implications for energy and climate policy as discussed further below.

Impacts of higher oil prices on tourism

In line with decision-making models discussed in the tourism literature (e.g. Woodside and King, 2001), the framework of how oil prices relate to tourism developed by Becken *et al.* (2010a) is hierarchical in nature. As can be seen in Figure 8.7, there are three different levels of impact: macroeconomic (i.e. global), destination choice and in-country.

As discussed in Becken (2011a), the most influential impacts on global tourism are likely to be at the macro-level. This is due to the impact higher oil prices have on the economic activity of oil-importing countries, which has flow-on effects that create lower incomes for consumers relative to a scenario in which oil prices are lower. Tourism, especially to long-haul destinations, is income elastic and a redistribution of global tourism flows is likely (for more detail on elasticities see Schiff and Becken, 2011). Higher oil prices also influence behaviour at the destination. For example, tourists may choose different transport modes, reduce travel distance or stay at different types of accommodation. Likewise, tourism operators

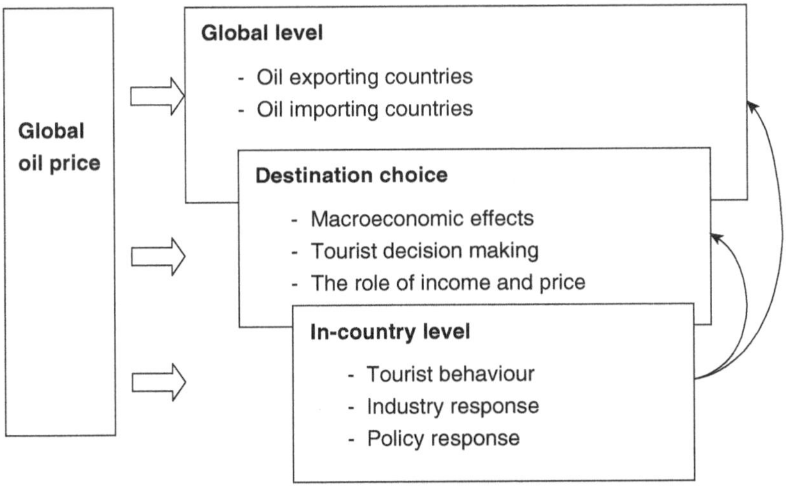

FIGURE 8.7 Hierarchical framework for oil price increase impacts on tourism. (Becken *et al.*, 2011a)

are more likely to invest in alternative energy sources and reduce dependency on (costly) oil. Interestingly, to date the price and availability of oil has been very rarely studied by tourism researchers, and it is also not explicitly included in tourism forecasts and scenario analyses. An exception is Draper *et al.* (2009), who anchored their tourism scenarios for the United Kingdom around energy and climate constraints.

In his article on New Tourism in a New Society, Leigh (2011) postulates the following changes for tourism:

- Any tourism types with large energy requirements will be very expensive and scarce. While there may be esoteric, luxurious facilities for the super-rich, who will be able to afford air travel, for most people air travel and sumptuous holiday facilities will be out of reach.
- Energy guzzling air travel will be greatly curtailed due to its high cost and so only available for those with large budgets.
- Most people will use the more efficient popular transport modes of train, bus and ship for intercontinental travel, although sea travel may be quite expensive too.
- There will be a high demand for local tourism – for the majority a local holiday at a local beach or mountain resort, for example, could be in reach.
- Popular tourism will have to be simpler, less luxurious and more localized, to cut down on the total price and on expensive consumption.
- There could be an opportunity for sustainable tourism to become more serious, beyond lip-service and window-dressing.
- Mass, highly commercialized tourism will decline – a smaller, localized tourism industry will (re)appear.
- Generally people will travel less and enjoy tourism which has simpler, cheaper requirements.
- There may be a revival for the true soul of tourism, hospitality, to begin to overshadow and replace amenities. The latter became the main focus of luxurious consumption-oriented tourism of the recent energy-hungry decades.

Clearly, reduced availability of oil or higher prices will have significant implications for tourist destinations (especially more remote ones, Plate 8.5), businesses, travellers and, as a result, for society as a whole (Becken, 2011a). The public and academic debate on this issue so far fails to understand the severity of the challenge.

Relationship between energy and climate change policies

Policies that address 'peak oil' often fall under the mantle of energy security policies. Energy security is the 'uninterrupted physical availability [of energy] at a price which is affordable, while respecting environment concerns' (International Energy Agency, 2010). Climate change policies (those directed at mitigation), in turn,

PLATE 8.5 Aviation fuel pump in a regional airport in Germany. (S. Becken)

aim at reducing the concentration of GHGs in the atmosphere. At first glance, the discussion of peak oil gives the impression of a solution to climate change in that the emissions of CO_2 are naturally declining over time. However, as Verbruggen and Al Marchohi (2010) point out, this is not necessarily the case. The reduced availability of cheap oil can provide investors with the confidence to finance activities such as the exploitation of coal and non-conventional resources as a substitute for conventional oil products. This would likely result in a large increase in GHG emissions. Policies directed at 'decarbonizing' economies are aimed at achieving energy security and mitigation goals.

There are only a few tourism-specific energy policies or initiatives, with policies related to climate change mitigation largely focused on carbon taxes and/or emissions trading schemes. While this reflects the generally greater emphasis on climate change compared with 'peak oil', the political efforts of tourism players run the risk of 'missing their mark'. As Becken (2011b) points out, the impacts of a carbon tax per barrel of oil (e.g. in the order of US$8 when a carbon price of US$20 is assumed) are miniscule compared with the recent price increases of aviation fuel as a result of the high global oil prices. The lack of political and practical engagement in the response to high oil prices, and also to anticipating future implications, is therefore astonishing.

A small number of examples of tourism–oil–climate interactions are presented overleaf.

Example	*Description*	*Source*
Climate Change and Peak Oil Strategy, Sunshine Coast Australia	The goal of the Climate Change and Peak Oil Strategy is to guide the transition to a low carbon, low oil, and resilient future for the Sunshine Coast (see also Chapters 3 and 5).	www.sunshinecoast.qld.gov.au/sitePage.cfm?code=cc-strategy
Oil vulnerability assessment tool, New Zealand	Online tool designed for businesses to assess their vulnerability to increasing oil prices in New Zealand.	www.lincoln.ac.nz/Research-Centres/LEaP/Climate-Change-Peak-Oil--Society/Projects/Tourism--Oil/
IATA	IATA treats energy security and climate change as a key policy issue. Analysis on 'The Impact of the Financial Crisis on Energy Investment' is available on IATA's website.	www.iata.org/whatwedo/Documents/economics/Birol_Energy_Investment.pdf
Diversification, Oman	Oman is seeking diversification of the economy away from oil to sectors such as tourism. The government is aiming to raise the contribution the hospitality industry makes to the Sultanate's gross domestic product (GDP) to around 3% by 2020.	www.omaninfo.com/tourism-and-travel-experinces/oman-doing-enough-put-itself-tourist-map.asp

Conclusion

The availability of cheap oil is critical for tourism as we know it. 'It is no longer a case of 'if' or 'when' there will be a plateau and eventual peak in world oil production. Rather, it is now a question of how we can prepare for a world with less oil, one in which energy security and climate policy play a dominant role (Hughes and Rudolph, 2011). The combination of the physical realities of a limited oil supply, and the political and moral (and also physical) need to reduce GHG emissions, come together to pose a major challenge for tourism that – it appears – has not been fully recognized by those responsible for providing leadership and guidance to the tourism industry at the global level.

9

CONCLUSION

The goal of this book is to provide an assessment of the most recent knowledge on the many interactions between climate change and tourism. This is accomplished by presenting and discussing a wide range of practical case studies and examples that highlight the many efforts already being made by the tourism sector to slow the rate of climate change, to reduce its adverse impacts, to take advantage of any benefits that might arise and to enhance the sector's resilience to future changes. The case studies were selected to illustrate climate change initiatives at both the policy and more practical levels, including actions implemented by tourism businesses and organizations. The book covers mitigation and adaptation, both separately and, more innovatively, by taking an integrated approach. It draws on lessons learned from all over the world.

The last decade has shown that many components of the climate system are changing near or even beyond the upper ranges of the IPCC projections made in its Fourth Assessment Report (IPCC, 2007a) and in its more recent Special Report on Extremes (IPCC, 2011). The decade 2001–2010 was the warmest on record, with average temperatures being 0.46°C above the 1961–1990 mean and 0.21°C warmer than the previous record decade, 1991–2000. Antarctic ice shelves are collapsing more rapidly than expected and the surface melting of the Greenland Ice Sheet is accelerating, causing concern for greater and faster-than-anticipated sea-level rise. Summer-time melting of Arctic sea ice has also accelerated, and is now exceeding the worst-case scenarios of the IPCC projections. The science highlights the need for urgent and effective action. In contrast, the political discussions on how best to address climate change progress at a pre-global warming glacial pace, or worse.

Against this background of urgency, and possibly because it is an interesting and novel field of research and application, knowledge on the linkages between climate change and tourism has increased substantially in the last five years. As shown in Chapter 2 (Recent Advances in Knowledge and Understanding), a large number of micro-level analyses provide detailed insights. For example, many studies have analysed the 'carbon footprints' of specific countries, tourism sub-sectors and activities. Research has also explored how to reduce GHGs emitted by the tourism sector, for example by the accommodation sub-sector, or by changes in travel behaviour such as 'slow tourism'. Past research has been dominated by energy audits and determining how GHG emissions might be reduced, as well as by studies covering such topics as the effect of tourist behaviour on emissions and perceptions of carbon offsetting. Recent years have seen an increasing number of studies seeking to understand when and how changes in climate will impact on tourism flows, tourist destinations and on tourist operations and activities. These have largely focused on alpine tourism (most notably skiing) and on beach tourism. But emissions and mitigation related to other types of tourism destinations and activities are now being studied, including World Heritage Sites and cruise ship tourism. While our understanding of, and experience in, climate change adaptation is still somewhat limited, there is an increasing number of technical studies that assess the efficiency and effectiveness of adaptation measures, for example in relation to water efficiency and conservation by the hotel industry.

One of the most significant gaps in knowledge and practice relates to our limited understanding of what constitutes sound climate change policy for tourism. Only a small number of studies have explored how policies are developed, implemented and evaluated with regard to either mitigation or adaptation in the tourism sector. A recent report (OECD and UNEP, 2011) provides an overview of the types of tourism policies that have been implemented by governments in OECD countries. However, detailed analyses and evaluations are lacking.

A promising avenue of research relates to future thinking, with a focus on climate change. A number of studies were identified that explore different futures, using scenario analysis or fiction writing to understand better how tourism might evolve over the next 20 or 50 years in a changing climate. Such analyses are invaluable for deciding on strategic responses and undertaking informed planning.

The book contains a large number of case studies that illustrate initiatives taken by tourism stakeholders to reduce tourism's contribution to climate change, to reduce the consequences of climate change for the sector, or to take advantage of the potential benefits provided by climate change. It was relatively easy to identify mitigation case studies, possibly reflecting a longer tradition of environmental management and of tourism seeking energy savings. Often, carbon dioxide emissions have been included as indicators or areas of action in wider environmental portfolios, thus aligning well with existing institutional arrangements and infrastructure. In contrast, as a newer dimension of tourism, case studies of both adaptation policy and practice are fewer in number, as well as in terms of the depth and rigour of analysis. This is especially so for adaptation that exploits the benefits of climate change. While this imbalance between mitigation and adaptation extends beyond the tourism sector, it is somewhat more pronounced for tourism because of the relative immaturity and the wider lack of long-term thinking in the tourism sector. Also, for some policy makers, planners and operators there is a lack of urgency. Observable impacts that call for an immediate response are still not fully apparent, though this situation is changing rapidly. However, both the increasing consequences of extreme weather events for tourism, and a growing interest in tourism disaster management, indicate that tourism is beginning to be more proactive in addressing weather extremes, climate variability and anticipated changes. The sector is becoming increasingly involved in addressing these hazards as part of a wider risk management approach.

Chapter 6 (Mitigation Practice) provides examples of good practice and concrete initiatives that seek to reduce GHG emissions. In contrast, Chapter 5 (Mitigation Policy) highlights that examples of policy initiatives are somewhat less abundant. This may indicate that many mitigation measures are implemented simply as part of regular business operations rather than being policy driven, in part because of savings if energy consumption is reduced, and because reducing emissions has been a popular theme over the last ten years or more. This ad hoc approach to mitigation is not necessarily detrimental, but mitigation without a sound policy framework, or at least an underlying strategy, risks being imperfect, partial and potentially counterproductive. Chapter 6 has shown that countries, destinations, organizations and businesses which have invested in a more strategic approach to climate change mitigation are more likely to reduce GHG emissions in an effective and lasting way, without compromising other business or development goals.

We question whether the global tourism sector can deliver on its 'aspirational' GHG emission reduction targets. This issue has been examined in even more detail by Scott *et al.* (2010), who argued that achieving the proposed emission reductions in tourism is largely dependent on emission reductions related to air travel. Such a result seems increasingly unlikely, given that an absolute reduction in emissions by the aviation sector is unlikely to be achieved before 2035 approximately. We observe that much of the tourism sector remains unaware of the implications of emission reduction targets proposed by a number of tourism players. The many and diverse consequences of the unavoidable transition towards a low carbon tourism economy need far greater attention by tourism stakeholders and destinations that depend on tourism.

The greater availability of adaptation practice case studies (see Chapter 4, Adaptation Practice), compared to those for policies and plans, has also been highlighted. It has been possible to identify only a few examples of tourism adaptation policies and plans that have already been implemented and can hence be judged for their effectiveness. As a result, the analysis in Chapter 3 (Adaptation Policy, Planning and Institutions) takes a slightly more theoretical approach. It is informative to note, though, that many countries, organizations and operators who have developed a comprehensive climate change policy are taking an integrated approach by including both adaptation and mitigation. Climate change policies for the Sunshine Coast, the Great Barrier Reef and the European Union highlight this. Reference to them in more than one chapter of this book (including Chapter 8, Integrated Responses to Climate Change) is testimony to their integrated approach.

An important element of both mitigation and adaptation is the monitoring of the effectiveness and performance of these responses to climate change, and the learning that results from this. In the case of mitigation, for example, the measurement of GHG emissions is a key factor in ensuring ongoing reductions in the carbon footprint. The monitoring and evaluation of adaptation is in its infancy, and not only for the tourism sector. The challenges for evaluating adaptation measures in the tourism sector have been discussed in Chapter 7 (Assessing Performance, and Learning) and examples of how these might be addressed have been provided.

For both categories of climate change response, the evaluation of individual policies or measures has not been systematically explored in tourism research and practice. Ongoing assessment of the effectiveness of both mitigation and adaptation by the tourism sector is important, no more so than when financial resources are limited, and governments and other stakeholders face many challenges in addition to climate change. Developing policies, plans and work programmes that address more than one issue is therefore critical. Chapter 8 (Integrated Responses to Climate Change) highlights the complementarity of climate change responses with those related to sustainable development, disaster risk reduction and energy security. The mainstreaming of tourism and climate change considerations in these policy areas, and vice versa, is shown to provide mutual benefits.

Our book highlights that, overall, the tourism sector has made considerable progress in addressing climate change, in developing more knowledge and a greater understanding, and in forming new or using existing institutions to create a common voice. When considered collectively, there are a large number of good ideas, innovative approaches and informative examples that showcase how climate change measures can be incorporated into tourism businesses or destination management. On the other hand, some case studies and examples presented in this book demonstrate continuing deficiencies in tourism's approach to climate change. Many of these examples do provide insights on where progress can be made. Also, relevant case studies could not always be found, such as for adaptation being undertaken by even some of the more common tourism businesses, and for some regions such as Africa and Latin America. It is to be hoped that these shortages will be rectified in the next few years.

The continuing growth of GHG emissions from the tourism sector as a whole indicates that much more needs to be done. Even though the emissions are decreasing in terms of tourist kilometre travelled and for many individual operations, these efforts are being more than offset by the overall growth of tourism. Moreover, the lack of concerted and effective action by key organizations such as the United Nations World Tourism Organisation (UNWTO) and the World Travel and Tourism Council (WTTC), shows that agreements, policies and institutional arrangements are not as effective as is desirable. Growing emissions from the international aviation sector and also maritime shipping – i.e. cruise ships – are pertinent examples.

Several recommendations arising from the findings in this book can be directed to the tourism sector. To improve credibility and enable learning, a uniform approach to carbon footprinting is essential. A suitable candidate is the approach promoted by the International Tourism Partnership and the WTTC. Developing robust, yet simple, mitigation technologies specifically for tourism will help accelerate emission reductions. Greater dissemination of information on these technologies and their economic and technical performance will be welcomed by tourism stakeholders. Similarly, there needs to be improved understanding of why adaptation is required, what the costs of not adapting to climate change are, and what adaptation measures are suitable for use by the tourism sector in specific situations and circumstances. This information is required at all levels of tourism, ranging from businesses, to destinations, countries and regions.

While knowledge and understanding of the relationships between tourism and climate change is increasing rapidly, greater linkages with other fields and sectors would be beneficial. Already there is increasing interest by non-tourism experts, for example engineers, in assisting the tourism sector to mitigate and adapt. Greater ties between the energy and tourism sector, as advocated by the International Centre of Peace through Tourism, are an example of such partnerships. Other important sectors for greater integration include agriculture, water and health. Some of these have a much longer history than tourism in terms of managing climate risks and opportunities, adapting to changes and using resources efficiently.

Finally, it is important to improve our understanding of the implications of climate change, both in terms of climatic impacts and the effects of climate change policies, on developing countries. An ethics-based approach to tourism management might help to decide which destinations truly benefit from tourism, which ones could use tourism as an adaptation tool (to diversify away from more traditional sectors) and which places might consider reducing their reliance on tourism for situations where it is more vulnerable than other sectors. More case studies from destinations in developing countries would be extremely useful in improving our understanding of the dynamics and dependencies such places have for tourism and the climate.

NOTES

4 Adaptation Practice

1 See: www.starwoodhotels.com/sheraton/property/overview/index.html?propertyID=485

5 Mitigation Policy

1 See also: www.eurostar.com/UK/uk/leisure/about_eurostar/environment/35_cut.jsp
2 For an example, see: www.nelsonnz.com/sustainable.tourism/
3 See the C2ES website: http://www.pewclimate.org/policy_center/international_policy).
4 See: http://hawaii.gov/dbedt/info/energy/policy/
5 See: http://climate-l.org/news/eu-considers-new-policy-on-tourism/
6 See: http://ec.europa.eu/environment/climat/aviation_en.htm
7 A detailed project plan aimed at donors is available at: http://idbdocs.iadb.org/wsdocs/getdocument.aspx?docnum=1866931

6 Mitigation Practice

1 More information on the city's actions is available at: www.vaxjo.se. Tourism-relevant examples can be found on the Växjö Alvesta tourist centre website: www.vaxjo.se/default.aspx?id=32457
2 See: www.airmauritius.com/csr.htm
3 Available at: www.enviro.aero/TheHold.aspx
4 See: http://destinationcarbonneutral.co.nz/intercity-s-track-record/
5 See: www.whistlerblackcomb.com
6 See: www.intrepidtravel.com/ourtrips/rt/climatechange.php

7 Assessing Performance, and Learning

1 http://carbon.trx.com/
2 www2.icao.int/en/carbonoffset/Pages/default.aspx
3 www.crctourism.com.au/Page/Tools+and+Products/Carbon+Calculator.aspx

4 www.carbonzero.co.nz/EmissionsCalc/tourismeditor.aspx
5 http://pata.calculator.earthcheck.org/
6 www.carbonfootprint.com/businesscarboncalculator.html
7 http://hes.e-benchmarking.org/index.html
8 See: http://energiekampagne-gastgewerbe.de
9 See: www.unwto.org/climate/index.php

REFERENCES

Accor (2010) Sustainable Development 2009/2010. Available at (12/02/2011) www.accor.com

ADB (2010) A Note on Climate Proofing Investment in the Transport Sector: Road Infrastructure Projects. Asian Development Bank (ADB), Manila, Philippines, 60pp.

ADFC (2010) Bett+Bike – Fahrradfreundliche Gastbetriebe. Homepage. Available at (12/01/2011) www.bettundbike.de

Agrawala, S. (2007) Climate Change in the European Alps: Adapting Winter Tourism and Natural Hazard Management. Organisation for Economic Cooperation and Development Publications, Paris, 136pp.

Aguilo, E., Riera, A. and Rossello, J. (2005) The Short-term Price Effect of a Tourist Tax through a Dynamic Demand Model. The Case of the Balearic Islands. *Tourism Management*, 26, 359–365.

Allison, I., *et al.* (2009) The Copenhagen Diagnosis: Updating the World on the Latest Climate Science. The University of New South Wales, Climate Change Research Centre (CCRC), Sydney, Australia, 60pp.

Amelung, B., Nicholls, S. and Viner, D. (2007) Implications of Global Climate Change for Tourism Flows and Seasonality. *Journal of Travel Research*, 45, 285–296.

Anger, A. and Koehler, J. (2010) Including Aviation Emissions in the EU ETS: Much Ado About Nothing? A Review. *Transport Policy*, 17, 38–46.

Angus Reid (2011) Britons Question Global Warming more than Americans and Canadians. Angus Reid Public Opinion, 6pp.

APEC International Centre for Sustainable Tourism (AICST) (2006) Tourism Risk Management: An Authoritative Guide to Managing Crises in Tourism. Available at (20/02/2011) www.apec.org

Archer, J. (2008) Flying Three Times Greener than Cruising. *The Telegraph*, Comment, 19 January. Available at (12/01/2011) www.telegraph.co.uk/travel/hubs/greentravel/739287/Flying-three-times-greener-than-cruising.html

Arkell, B. P. Darch, G. J. C. (2006) Impact of Climate Change on London's Transport Network. *Municipal Engineer*, 159, 231–237.

Arndt, D. S., Baringer, M. O. and Johnson, M. R. (Eds.) (2010) State of the Climate in 2009. *Bulletin of the American Meteorological Society*, 91(7), S1–S224.

Arnell, N. W. (2010) Adapting to Climate Change: An Evolving Research Programme. *Climatic Change*, 100, 107–111.

ASPIRE (2010) Homepage. Available at (05/01/2011) www.aspire-green.com/about/default.asp

Atmosfair (2009) Annual Report 2009. Available at (25/01/2011) www.atmosfair.de/fileadmin/user_upload/Medien/Jahresbericht_2009_EN_Website.pdf

Baker, A. (2011) How Can Air Travel Contribute to the costs of Adapting to Climate Change? The Feasibility of the International Air Passenger Adaptation Levy (IAPAL) as a Market Governance Mechanism. International Institute for Environment and Development, London, United Kingdom, 44pp.

Baker, C. J., Chapman, L., Quinn, A. and Dobney, K. (2010) Climate Change and the Railway Industry: A Review. *Journal of Mechanical Engineering Science*, 224(C), Special Issue 519.

Ballou, P., Chen, H. and Horner, J. D. (2008) Advanced Methods of Optimizing Ship Operations to Reduce Emissions Detrimental to Climate Change. *Oceans*, 15–18 September, 1–12.

Bank, M. and Wiesner, R. (2011) Determinants of Weather Derivatives Usage in the Austrian Winter Tourism Industry. *Tourism Management*, 32(1), 62–68.

Bark, R. H., Colby, B. G. and Dominguez, F. (2010) Snow Days? Snowmaking Adaptation and the Future of Low Latitude, High Elevation Skiing in Arizona, USA. *Climatic Change*, 102(3–4), 467–491.

Barnett, J. and Campbell, J. (2010) *Climate Change and Small Island States: Power, Knowledge, and the South Pacific*. London: Earthscan.

Barnett, J. and O'Neill, S. J. (2010) Maladaptation. *Global Environmental Change*, 20, 211–213.

Barr, S., Shaw, G., Coles, T. and Prillwitz, J. (2010) A Holiday is a Holiday: Practicing Sustainability, Home and Away. *Journal of Transport Geography*, 18, 474–481.

Becken, S. (2004) Climate Change and Tourism in Fiji: Vulnerability, Adaptation and Mitigation. Final Report. University of the South Pacific, Suva, Fiji.

Becken, S. (2007) Tourists' Perception of International Air Travel's Impact on the Global Climate and Potential Climate Change Policies. *Journal of Sustainable Tourism*, 15(4), 351–368.

Becken, S. (2008a) Indicators for Managing Tourism in the Face of Peak Oil. *Tourism Management*, 29, 695–705.

Becken, S. (2008b) *The Carbon Footprint of Domestic Tourism in New Zealand*. Wellington, New Zealand. Hikurangi Foundation.

Becken, S. (2009) Global Challenges for Tourism and Transport: How will Climate Change and Energy affect the Future of Tourist Travel? Chapter 8 (pp. 328–342). In S. Page. *Transport and Tourism – Global Perspectives*. Third edition. Essex: Pearson Education Limited.

Becken, S. (2010) The Importance of Climate and Weather for Tourism: Literature Review. LEaP Research Paper, Centre for Land, Environment and People (LEaP), Lincoln University, Lincoln, New Zealand, 23pp.

Becken, S. (2011a) A Critical Review of Tourism and Oil. *Annals of Tourism Research*, 38(2), 359–379.

Becken, S. (2011b) Responsible Aviation and Peak Oil. Presented at Responsible Aviation Conference, Manchester, UK, 17 June 2011. Available at www.lincoln.ac.nz/leap

Becken, S. (in press) Operators' Perceptions of Energy Use and Actual Saving Opportunities for Tourism Accommodation. *Pacific Tourism Review*, accepted.

Becken, S. and Clapcott, R. (2011) Developing Public Policy for Climate Change in the Tourism Sector. *Journal of Policy Research in Tourism, Leisure and Events*, 3(1), 1–17.

Becken, S., Frampton, C. and Simmons, D. (2001) Energy Consumption Patterns in the Accommodation Sector – the New Zealand case. *Ecological Economics*, 39(3), 371–386.

Becken, S. and Hay, J. (2007) *Tourism and Climate Change: Risks and Opportunities*. Clevedon: Channel View Publications.

Becken, S. and Hughey, K. (submitted) Tourism and Natural Disaster Risk Reduction – Opportunities for Integration. *Tourism Management*.

Becken, S. and Patterson, M. (2006) Measuring National Carbon Dioxide Emissions from Tourism as a Key Step Towards Achieving Sustainable Tourism. *Journal of Sustainable Tourism*, 14(4), 323–338.

Becken, S., Wilson, J., Hendrikx, J. and Hughey, K. (2010a) The Impact of Climate Variability on Tourism Businesses in Wanaka and Queenstown. LEaP Research Paper, Centre for Land, Environment and People (LEaP), Lincoln University, Lincoln, New Zealand, 9pp.

Becken, S., Wilson, J. and Hughey, K. (2011) Planning for Climate, Weather and Other Natural Disasters – Tourism in Northland. LEaP Research Paper 1. Available at (16/11/2011) www.lincoln.ac.nz/leap

Becken, S., Wilson, J. and Reisinger, A. (2010b) Tourism, Climate and Weather – a New Zealand Perspective. LEaP Report 20. Available at (02/08/2011) www.lincoln.ac.nz/leap

Beniston, M., Uhlmann, B., Goyettea, S. and Lopez-Morenob, J. I. (2011) Will Snow-abundant Winters Still Exist in the Swiss Alps in an Enhanced Greenhouse Climate? *International Journal of Climatology*, 31(9), 1257–1263.

Berrang-Ford, L., Ford, J. and Patterson, J. (2011) Are we Adapting to Climate Change? *Global Environmental Change*, 21, 25–33.

Bicknell, S. and McManus, P. (2006) The Canary in the Coalmine: Australian Ski Resorts and their Response to Climate Change. *Geographical Research*, 44(4), 386–400.

Bigano, A., Bosello, F., Roson, R. and Tol, R. S. J. (2008) Economy-wide Impacts of Climate Change: A Joint Analysis for Sea Level Rise and Tourism. *Mitigation and Adaptation Strategies for Global Change*, 13(8), 765–791.

Bilimoria, K., Sridhar, B., Chatterji, G., Sheth, K. and Grabbe, S. (2000) FACET: Future ATM Concepts Evaluation Tool. 3rd USA/Europe Air Traffic Management RandD Seminar, Napoli, Italy, 9pp.

Bird, E. C. F. (2000) *Coastal Geomorphology: An Introduction*. Chichester: John Wiley.

Bode, S., Hapke, J. and Zisler, S. (2003) Need and Options for a Regenerative Energy Supply in Holiday Facilities. *Tourism Management*, 24, 257–266.

Bohdanowicz, P. (2006) Environmental Awareness and Initiatives in the Swedish and Polish Hotel Industries – Survey Results. *International Journal of Hospitality Management*, 25, 662–682.

Bohdanowicz, P., Churie-Kallhauge, A. and Martinac, I. (2001) Energy-Efficiency and Conservation in Hotels – Towards Sustainable Tourism. 4th International Symposium on Asia Pacific Architecture, Hawai'i, April 2001.

Bohdanowicz, P. and Martinac, I. (2002) Thermal Comfort and Energy Saving in the Hotel Industry. Paper 16A.8. Available at (08/02/2012) www.greenthehotels.com/eng/BohdanowiczMartinacKansas2002.pdf

Bohdanowicz, P. and Martinac, I. (2007) Determinants and Benchmarking of Resource Consumption in Hotels – Case Study of Hilton International and Scandic in Europe. *Energy and Buildings*, 39, 82–95.

Bomhard, B. and Midgley, G. (2005) Securing Protected Areas in the Face of Global Change: Lessons Learned from the South African Cape Floristic Region. A Report by the Ecosystems, Protected Areas, and People Project. IUCN, Bangkok and SANBI, Cape

Town. Available at (08/02/2012) http://iucn.org/about/union/commissions/wcpa/wcpa_puball/wcpa_pubsubject/wcpa_climatepub/?2086/Securing-Protected-Areas-in-the-Face-of-Global-Change-Lessons-learned-from-the-South-African-Cape-Floristic-Region

Boucher, O. (2011) Seeing Through Contrails. *Nature Climate Change*, 1, 24–25. Available at (21/07/2011) www.nature.com/natureclimatechange

Bows, A. (2010) Aviation and Climate Change: Confronting the Challenge. *The Aeronautical Journal*, 114(1158), 459–468.

Bows, A., Anderson, K. and Peeters, P. (2009) Air Transport, Climate Change and Tourism. *Tourism and Hospitality Planning and Development*, 6(1), 7–20.

Bristow, A. L., Tight, M., Pridmore, A. and May, A. D. (2008) Developing Pathways to Low Carbon Land-based Passenger Transport in Great Britain by 2050. *Energy Policy*, 36, 3427–3435.

British Department of Transport (2008) Aviation Emissions Cost Assessment 2008. Available at (18/01/2011) www.dft.gov.uk/pgr/aviation/environmentalissues/aviationemissions-costassess/aviationemissionscost.pdf

Brons, M., Pels, E., Nijkamp, P. and Rietveld, P. (2002) Price Elasticities of Demand for Passenger Air Travel: A Meta-analysis. *Journal of Air Transport Management*, 8, 165–175.

Brouwer, R., Brander, L. and Beukering, P. Van (2008) A Convenient Truth: Air Travel Passengers' Willingness to Pay to Offset their CO_2 Emissions. *Climatic Change*, 90, 299–313.

Bruno, P. (2011) The Future of Arctic Shipping Operations: Expanding Arctic Shipping Challenges Commercial and Leisure Vessels. Available at (08/02/2012) http://maritime.about.com/od/Climate/a/The-Future-Of-Arctic-Shipping-Operations.htm

Butler, R. W. (1980) The Concept of a Tourist Area Cycle of Evolution: Implications for Management of Resources. *Canadian Geographer*, 24, 5–12.

Calbó, J. (2010) Possible Climate Change Scenarios with Specific Reference to Mediterranean Regions. In S. Sabater and D. Barcelo (Eds.), Water Scarcity in the Mediterranean: Perspectives Under Global Change, *Handbook of Environmental Chemistry*, 8, 1–13.

Caribbean Community Climate Change Centre (2009) Climate Change and the Caribbean: A Regional Framework for Achieving Development Resilient to Climate Change (2009–2015). Caribbean Community Climate Change Centre (CCCCC), 41pp.

Chan, W. W. and Lam, J. C. (2003) Energy-saving Supporting Tourism: A Case Study of Hotel Swimming Pool Heat Pump. *Journal of Sustainable Tourism*, 11(1), 74–83.

Chan, W. W., Mak, L. M., Chen, Y. M., Wang, Y. H., Xie, H. R., Hou, G. Q. and Li, D. (2008) Energy Saving and Tourism Sustainability: Solar Control Window Film in Hotel Rooms. *Journal of Sustainable Tourism*, 16(5), 563–574.

Charara, N., Cashman, A., Bonnell, R. and Gehr, R. (2011) Water Use Efficiency in the Hotel Sector of Barbados. *Journal of Sustainable Tourism*, 19(2), 231–245.

Chester, M. and Horvath, A. (2009) Environmental Assessment of Passenger Transportation Should Include Infrastructure and Supply Chains. *Environmental Research Letters*, 4, 1–8.

Chomhairle Oidhreachta/The Heritage Council (2009) Climate Change, Heritage and Tourism – Implications for Ireland's Coast and Inland Waterways. Summary Document, The Heritage Council of Ireland Series, 14pp.

Cioccio, L. and Michael, E. J. (2007) Hazard or Disaster: Tourism Management for the Inevitable in Northeast Victoria. *Tourism Management*, 28, 1–11.

Clean Air – Cool Planet (2006) A Consumers' Guide to Retail Carbon Offset Providers. Available at (08/02/2012) www.cleanair-coolplanet.org/ConsumersGuidetoCarbonOffsets.pdf

Climate Commission (2011) The Critical Decade. Climate Science, Risks and Responses. Climate Commission Secretariat, Department of Climate Change and Energy Efficiency, Commonwealth of Australia, Canberra, Australia, 72pp.

Climate Southwest (2010) Preparing for Change: Climate-proof Your Tourism Business – Online Toolkit. Available at (08/02/2012) www.oursouthwest.com/climate/sr-tourism.htm

Coghlan, A. and Prideaux, B. (2009) Welcome to the Wet Tropics: The Importance of Weather in Reef Tourism Resilience. *Current Issues in Tourism*, 12(2), 89–104.

Commission for Environmental Cooperation (no date) Whistler, British Columbia, Canada – Integrated Energy, Air Quality and Greenhouse Gas Management Plan. Available at (16/11/2011) http://cec.org/municipalenergy/case_studies.asp?xmlFile=whistlerandvarlan=en

Commission of the European Communities (2009) Adapting to Climate Change: Towards a European Framework for Action. White Paper COM (2009) 147 Final, European Commission, Brussels, 17pp. Available at (05/02/2012) http://eur-lex.europa.eu/LexUriServ/LexUriServ.do?uri=COM:2009:0147:FIN:EN:PDF

Commission of the European Communities (2010) Europe, the World's No. 1 Tourist Destination – A New Political Framework for Tourism in Europe, European Commission, Brussels, 14pp. Available at (05/02/2012) http://ec.europa.eu/enterprise/sectors/tourism/files/communications/communication2010_en.pdf

Commonwealth of Australia (2007) National Tourism Incident Response Plan. Available at (11/03/2011) www.ntirp.gov.au

Coombers, E. G., Jones, A. P. and Sutherland, W. (2008) The Biodiversity Implications of Changes in Coastal Tourism due to Climate Change. *Environmental Conservation*, 35(4), 319–330.

Coombers, E. G., Jones, A. P. and Sutherland, W. J. (2009) The Implications of Climate Change on Coastal Visitor Numbers: A Regional Analysis. *Journal of Coastal Research*, 25(4), 981–990.

Cowper-Smith, A. and de Grosbois, D. (2011) The Adoption of Corporate Social Responsibility Practices in the Airline Industry. *Journal of Sustainable Tourism*, 19(1), 59–78.

CSIRO (2007) Climate Change in Australia. Observed Changes and Projections. Summary. Available at (01/08/2010) www.climatechangeinaustralia.gov.au

Dalton, G. J., Lockington, D. A. and Baldock, T. E. (2007) A Survey of Tourist Operator Attitudes to Renewable Energy Supply in Queensland, Australia. *Renewable Energy*, 32(4), 567–586.

Dalton, G. J., Lockington, D. A. and Baldock, T. E. (2008) Feasibility Analysis of Stand-alone Renewable Energy Supply Options for a Large Hotel. *Renewable Energy*, 33, 1475–1490.

Davidson, D. and Kummer, B. (2008) Regulatory Report: FDA Regulation of Domestic Waterborne Transportation. *Food Safety Magazine*, April/May, 43–45. Available at (08/02/2012) www.foodsafetymag-digital.com/foodsafetymag/20080405?pg=4#pg45

Dawson, J., Scott, D. and McBoyle, G. (2009) Climate Change Analogue Analysis of Ski Tourism in the Northeastern USA. *Climate Research*, 39, 1–9.

Dawson, J., Stewart, E. J., Lemelin, H. and Scott, D. (2010) The Carbon Cost of Polar Bear Viewing Tourism in Churchill, Canada. Special Issue (edited by Scott, D. and Becken, S.): Tourism: Adapting to Climate Change and Climate Policy. *Journal of Sustainable Tourism*, 18(3), 319–336.

de Bruijn, K., Dirven, R., Eijgelaar, E. and Peeters, P. (2011) Travelling Large in 2009: The Carbon Footprint of Dutch Holidaymakers in 2009 and the Development Since 2002.

Breda, Netherlands: NHTV Breda University of Applied Sciences; NRIT Research; NBTC-NIPO Research.

DEFRA (2007) Public Understanding of Sustainable Leisure and Tourism. A research report completed for the Department for Environment, Food and Rural Affairs by the University of Surrey. Available at (20/03/08) www.defra.gov.uk

DEFRA (2010) Defra's Climate Change Plan. Department for Environment, Food and Rural Affairs (DEFRA), London, UK, 172pp.

de Freitas, C. R. (2003) Tourism Climatology: Evaluating Environmental Information for Decision Making and Business Planning in the Recreation and Tourism Sector. *International Journal of Biometeorology*, 48, 45–54.

Department of Resources, Energy and Tourism (2008) Tourism and Climate Change – A Framework for Action. Available at (12/12/2010) www.ret.gov.au/tourism/policy/Tourism%20and%20Climate%20Change/frameworkforaction/Pages/default.aspx

Department of Resources, Energy and Tourism (2009) Climate Change Guide. Mitigation and Adaptation Measures for Australian Tourism Operators. Canberra: Commonwealth of Australia. Available at (12/07/2011) www.ret.gov.au/tourism/Documents/Tourism%20and%20Climate%20Change/2141259A_TOURISM_CLIMATE_CHANGE_Workshop%20guide_v12.pdf

Des Bois, R. (2011) The State of the Arctic. Association de Protection de l'Homme et de l'Environnement, Paris, France, 3pp. Available at (05/02/2012) www.robindesbois.org/english/treaty_search_and_rescue_arctic.pdf

Dhakal, S. (2002) Role of Government, Private Sector and Civic Society in Promoting Battery Operated Electric Three-wheelers in Kathmandu, Nepal. Working paper, Urban Environmental Management Project, Institute for Global Environmental Strategies (IGES). Available at (18/01/2011) www.iges.or.jp/en/ue/pdf/dhakal/dhakal_EVKTM.pdf

Dickinson, J. and Lumsdon, L. (2010) *Slow Travel and Tourism*. London: Earthscan, 232pp.

Dieke, P. U. C. (2005) Critical Success Factors for Tourism Development in Less Developed Countries (LDCs). Paper presented at the Debate on Tourism in the Toolbox of Development Projects, University of Leuven, Belgium, October 2005, 20pp.

Dodds, R. (2010) Destination Marketing Organizations and Climate Change – The Need for Leadership and Education. *Sustainability*, 2, 3449–3464.

Dodds, R. and Graci, S. (2009) Canada's Tourism Industry – Mitigating the Effects of Climate Change: A Lot of Concern but Little Action. *Tourism and Hospitality Planning and Development*, 6(1), 39–51.

Dodds, R. and Kelman, I. (2008) How Climate Change is Considered in Sustainable Tourism Policies: A Case of the Mediterranean Islands of Malta and Mallorca. *Tourism Review International*, 12, 57–70.

Dolnicar, S. (2010) Identifying Tourists with Smaller Environmental Footprints. *Journal of Sustainable Tourism*, 18(6), 717–734.

Dolnicar, S., Laesser, C. and Matus, K. (2009) Short Haul City Travel is Truly Environmentally Sustainable Tourism Management. *Tourism Management*, 31(4), 505–512.

Dorfman, M. and Mehta, M. (2011) Thirsty for Answers: Preparing for the Water-related Impacts of Climate Change in American Cities. Natural Resources Defense Council, New York, US, 132pp.

Draper, S., Goodman, J., Hardyment, R. and Murray, V. (2009) Tourism 2023: Four Scenarios, a Vision and a Strategy for UK Outbound Travel and Tourism. Forum for the Future. Available at (12/10/2010) www.forumforthefuture.org/projects/travel-and-tourism

Draper, S. and Murray, V. (2008) Paradise Found - Guiding Principles for a Sustainable Tourism Development. Forum for the Future. Available at (19/04/2011) www.forumforthefuture.org/files/paradise-found-web.pdf

Dubois, G., Peeters, P., Ceron, J. P. and Goessling, S. (2011) The Future Tourism Mobility of the World Population: Emission Growth Versus Climate Policy. *Transportation Research Part A*, 45(10), 1031–1042.

Duffy-Mayers, L. (2010) Caribbean Hotel Energy Efficiency Action Program (CHENACT). Presented at CREF 2010, Atlantis Hotel Bahamas. Available at (13/03/2011) www.onecaribbean.org/content/files/CHENACT%20-CREF.pdf

Duvat, V. (2011) Interest of Quality-based Policies for Integrated Coastal Zone Management Implementation: Lessons Learnt from a French Case Study. *Ocean and Coastal Management*, 54(11), 841–853.

Dwyer, L., Forsyth, P. and Dwyer, W. (2011) *Tourism Economics and Policy*. Bristol: Channel View Publications, 776pp.

Dwyer, L., Forsyth, P. and Rao, P. (2001) PPPs and the Price Competitiveness of International Tourism Destinations. Joint World Bank/OECD Seminar on Purchasing Power Parities. Recent Advances in Methods and Applications, 30 January to 2 February, 2001, Washington.

Dwyer, L., Forsyth, P., Spurr, R. and Hoque, S. (2010) Estimating the Carbon Footprint of Australian Tourism. Special Issue (edited by Scott, D. and Becken, S.): Tourism: Adapting to Climate Change and Climate Policy. *Journal of Sustainable Tourism*, 18(3), 377–392.

EarthCheck (2010) Efficient Air Conditioning. EarthCheck Pty Ltd., Queensland, Australia, 3pp.

ECLAC (2011) The Economics of Climate Change in the Caribbean – Summary Report 2011. United Nations Economic Commission For Latin America and the Caribbean (ECLAC), Subregional Headquarters for the Caribbean, Port-of-Spain, Trinidad and Tobago, 106pp.

Ehmer, P. and Heymann, E. (2008) Klimawandel und Tourismus: Wohin geht die Reise? Aktuelle Themen, 416. Frankfurt: Deutsche Bank Research, 30pp.

Eijgelaar, E., Thaper, C. and Peeters, P. (2010) Antarctic Cruise Tourism: The Paradoxes of Ambassadorship, 'Last Chance Tourism' and Greenhouse Gas Emissions. Special Issue (edited by Scott, D. and Becken, S.): Tourism: Adapting to Climate Change and Climate Policy. *Journal of Sustainable Tourism*, 18(3), 355–376.

Elliott, V. and Buckley, C. (2008) Fuel Prices Spark Holiday Crunch as Air Surcharges Soar. *The Times*. Available at (03/10/08) http://business.timesonline.co.uk/tol/business/industry_sectors/transport/article4029739.ece

Endler, C. and Matzarakis, A. (2010) Analysis of High-resolution Simulations for the Black Forest Region from a Point of View of Tourism Climatology – A Comparison between Two Regional Climate Models (REMO and CLM). *Theoretical and Applied Climatology*, 103(3–4), 427–440.

EnergyStar (no date) Hotels: An Overview of Energy Use and Energy Efficiency Opportunities. Available at (08/02/2012) www.energystar.gov/ia/business/challenge/learn_more/Hotel.pdf

Engelbrecht, F. A., McGregor, J. L. and Engelbrecht, C. J. (2009) Dynamics of the Conformal Cubic Atmospheric Model Projected Climate Change Signal over Southern Africa. *International Journal of Climatology*, 29, 1013–1033.

Eugenio-Martin, J. L. and Campos-Soria, J. A. (2009) Climate in the Region of Origin and Destination Choice in Outbound Tourism Demand. *Tourism Management*, 31(6), 744–753.

EUROCONTROL (2006) Delays to Air Transport in Europe Digest – Annual 2005. Available at (08/02/2012) www.eurocontrol.int/eCoda/codarep/2005/annual_report_2005.pdf

EUROCONTROL (2008) Challenges of Growth. Summary Report. Available at (05/01/2011) www.eurocontrol.int/statfor/gallery/content/public/documents/Challenges%20of%20Growth%2008%20Summary%20Report%20v1.pdf

European Commission (2010) FLAG factsheet – France – Pays Marennes Oléron. Available at (08/02/2012) https://webgate.ec.europa.eu/fpfis/cms/farnet/content/flag-factsheet-france-pays-marennes-oléron

European Commission (2011) Allocation of Aviation Allowances in an EEA-wide Emission Trading System. Press release. Available at (18/07/2011) http://ec.europa.eu/clima/news/articles/news_2011070102_en.htm

European Community (2009) Making Aviation Greener – A Snapshot on European Measures for Sustainable Transport. CAAF/09-IP/5. Presented at Conference on Aviation and Alternative Fuels, Rio de Janeiro, Brazil, 16–18 November 2009.

Executive Office (2011) Massachusetts Climate Change Adaptation Report. Executive Office of Energy and Environmental Affairs and the Adaptation Advisory Committee, Commonwealth of Massachusetts, Boston, Massachusetts, US, 129pp.

Falk, M. (2009) A Dynamic Panel Data Analysis of Snow Depth and Winter Tourism. *Tourism Management*, 31(6), 912–924.

FAO (2009) Climate Change: Implications for Food Safety. Food and Agriculture Organisation (FAO), Rome, Italy, 49pp.

FAO/WHO (undated) Assuring Food Safety and Quality: Guidelines for strengthening National Food Control Systems. Joint publication of the Food and Agriculture Organisation (FAO) and the World Health Organisation (WHO), Rome and Geneva, 76pp. Available at (20/03/2012) http://www.who.int/foodsafety/publications/capacity/en/English_Guidlines_Food_Control.pdf.

Faulkner, B. (2001) Towards a Framework for Tourism Disaster Management. *Tourism Management*, 22(2), 135–147.

Fenner, R. (2011) Queensland Flood Sets Back $4 Billion State Tourism Industry. 14 January 2011. Available at (07/06/2011) www.bloomberg.com/news/2011–01–14/queensland-s-tourism-industry-faces-ruin-as-floods-engulf-australian-state.html

Ferris, E. and Petz, D. (2011) A Year of Living Dangerously. A Review of Natural Disasters in 2010. April, 2011. London: The Brookings Institution, London School of Economics.

Filimonau, V., Dickinson, J. E., Robbins, D. and Reddy, V. (2011) A Critical Review of Methods for Tourism Climate Change Appraisal: Life Cycle Assessment as a New Approach. *Journal of Sustainable Tourism*, 19(3), 301–324.

Fischer, E. M. and Schär, C. (2010) Consistent Geographical Patterns of Changes in High-impact European Heatwaves. *Nature Geoscience*, 3, 398–403.

Forbes, D. L. (Ed.) (2011) State of the Arctic Coast 2010 – Scientific Review and Outlook. International Arctic Science Committee, Land–Ocean Interactions in the Coastal Zone, Arctic Monitoring and Assessment Programme, International Permafrost Association. Helmholtz-Zentrum, Geesthacht, Germany, 178 pp. Available at (08/02/2012) http://arcticcoasts.org

Forsythe, P., Hoque, S., Dwyer, L., Spurr, R., Van Ho, T. and Pambudi, D. (2008) Carbon Footprint of Australian Tourism. Sustainable Tourism CRC. Available at (11/04/2011) www.sustainabletourismonline.com/awms/Upload/Resource/bookshop/100072%20Forsyth%20Carbon_FootprintWEB.pdf

Friedrich, P., Blechinger, H. and Shah, K. U. (2011) A Multi-criteria Evaluation of Policy Instruments for Climate Change Mitigation in the Power Generation Sector of Trinidad and Tobago. *Energy Policy*, 39(10), 6331–6343.

Gardiner L., Firestone D., Waibl, G. *et al.* (2009) Climate Change Effects on the Land Transport Network Volume One: Literature Review and Gap Analysis. Report for the New Zealand Transport Agency Research, Wellington, 229pp.

GBRMPA (2009) The Great Barrier Reef Tourism Climate Change Action Strategy 2009–2012. Great Barrier Reef Marine Park Authority (GBRMPA), Townsville, Australia, 56pp.

GBRMPA (2010) Coral Bleaching Response Plan 2010–2011. Great Barrier Reef Marine Park Authority (GBRMPA), Townsville, Australia, 36pp.

GEF (2008) Elements for a Monitoring and Evaluation Framework for Climate Change Adaptation Projects: Lessons from GEF Climate Change Adaptation Projects. Prepared by the GEF Evaluation Office in Cooperation with the GEF Adaptation Task Force, Global Environment Facility (GEF), Washington, DC, 58pp.

Gittens, A. (2010) Airports and Climate Change Adaptation. ICAO Council Briefing, Montreal, 16 April 2010.

GIZ and WRI (2011) Making Adaptation Count: Concepts and Options for Monitoring and Evaluation of Climate Change Adaptation. Published by Deutsche Gesellschaft fuür Internationale Zusammenarbeit (GIZ) GmbH on behalf of Bundesministerium fuür wirtschaftliche Zusammenarbeit und Entwicklung (BMZ), Referat 312, Klimapolitik und Klimafinanzierung, Eschborn, Germany, 96pp.

Gleick, P. H. and Cooley, H. S. (2009) Energy Implications of Bottled Water. *Environmental Research Letters*, 4, 1–6.

Glick, P., Chmura, H. and Stein, B. A. (2011) Moving the Conservation Goalposts: A Review of Climate Change Adaptation Literature. Adaptation 2011, National Wildlife Federation and National Council for Science and the Environment, Washington, DC, 25pp.

Global Travel (2008) Airlines Fuel Surcharge Chart. Available at (03/10/08) www.global-travel.com.sg/airline-fuel.php

Gordijn, H. (2010) The German Aviation Tax – Will it Last? Airlines Magazine. Available at (12/09/2011) www.aerlines.nl/index.php/2010/the-german-aviation-tax-will-it-last/

Gössling, S. (2009) Carbon Neutral Destinations: A Conceptual Analysis. *Journal of Sustainable Tourism*, 17(1), 17–37.

Gössling, S. (2010) *Carbon Management in Tourism: Mitigating the Impacts on Climate Change.* Oxford: Routledge.

Gössling, S., Borgström Hansson, C., Hörstmeier, O. and Saggel, S. (2002) Ecological Footprint Analysis as a Tool to Assess Tourism Sustainability. *Ecological Economics*, 43(2–3), 199–211.

Gössling, S., Broderick, J., Upham, P., Ceron, J. P., Dubois, G., Peeters, P. and Strasdas, W. (2007) Voluntary Carbon Offsetting Schemes for Aviation: Efficiency, Credibility and Sustainable Tourism. *Journal of Sustainable Tourism*, 15(3), 223–248.

Gössling, S., Garrod, B., Aall, C., Hille, J. and Peeters, P. (2010b) Food Management in Tourism: Reducing Tourism's Carbon 'Foodprint'. *Tourism Management*, 32(3), 534–543.

Gössling, S. and Hall, M. C. (2008) Swedish Tourism and Climate Change Mitigation: An Emerging Conflict? *Scandinavian Journal of Hospitality and Tourism*, 8(2), 141–158.

Gössling, S., Hall, M., Peeters, P. and Scott, D. (2010a) The Future of Tourism: Can Tourism Growth and Climate Policy be Reconciled? A Mitigation Perspective. *Tourism Recreation Research*, 35(2), 119–130.

Gössling, S., Peeters, P. and Scott, D. (2008) Consequences of Climate Policy for International Tourist Arrivals in Developing Countries. *Third World Quarterly*, 29, 873–901.

Gössling, S. and Schuhmacher, K. (2010) Implementing Carbon Neutral Destination Policies: Issues from the Seychelles. *Journal of Sustainable Tourism*, 18(3), 377–391.

Gössling, S. and Upham, P. (2009) *Climate Change and Aviation*. London: Earthscan.

Government of Australia (2010) Kakadu National Park Climate Change Strategy 2010–2015. Director of National Parks, Department of the Environment, Water, Heritage and the Arts, Canberra, Australia, 18pp.

Government of Jamaica (2009) Vision 2030 Jamaica, National Development Plan. Planning Institute of Jamaica, Kingston, Jamaica, 353pp.

Government of Mexico (2007) National Climate Change Strategy. Interministerial Commission on Climate Change, Government of Mexico, Mexico City, Mexico.

Government of Mexico (2009) Special Climate Change Programme. Interministerial Commission on Climate Change, Government of Mexico, Mexico City, Mexico.

Government of Ontario (2011) Climate Ready: Ontario's Adaptation Strategy and Action Plan, 2011–2011. Government of Ontario, Toronto, Canada, 124pp.

Government of Saint Lucia (no date) Saint Lucia National Climate Change Adaptation Policy and Strategy. Ministry of Physical Development, Environment and Housing, Government of Saint Lucia, Castries, Saint Lucia, 35pp.

Government of the United Kingdom (2008) Climate Change Act (2008), 108pp. Available at (05/02/2012) www.legislation.gov.uk/ukpga/2008/27/contents

Greene, D. L., Hopson, J. L. and Li, J. (2006) Have We Run Out of Oil, Yet? Oil Peaking Analysis from an Optimist's Perspective. *Energy Policy*, 34, 515–531.

Greenhouse Gas Protocol Initiative, The (2010) Homepage. Available at (08/01/2011) www.ghgprotocol.org/about-ghgp

Hallegatte, S. (2009) Strategies to Adapt to an Uncertain Climate Change. *Global Environmental Change*, 19, 240–247.

Hamilton, J. M. and Tol, R. S. J. (2007) The Impact of Climate Change on Tourism in Germany, the UK and Ireland: A Simulation Study. *Regional Environmental Change*, 7(3), 161–172.

Harding, P. and Malone, R. (2005) Green Events Guide. A guide to reducing the environmental impacts of conferences and seminars. Available at (03/01/2011) www.oursouthwest.com/SusBus/greener-events.pdf

Hares, A., Dickinson, J. and Wilkes, K. (2010) Climate Change and the Air Travel Decisions of UK Tourists. *Journal of Transport Geography*, 18, 466–473.

Hartz, D. A., Brazel, A. J. and Heisler, G. M. (2006) A Case Study in Resort Climatology of Phoenix, Arizona, USA. *International Journal of Biometeorology*, 51, 73–83.

Hedger, M. M., Mitchell, T., Leavy, J., Greeley, M., Downie, A. and Horrocks, L. (2008) Desk Review: Evaluation of Adaptation to Climate Change from a Development Perspective. A study commissioned by the GEF Evaluation Office and financed by DfID, Institute of Development Studies, London, United Kingdom, 60pp.

Hein, L., Metzger, M. and Moreno, A. (2009) Potential Impacts of Climate Change on Tourism: A Case Study for Spain. *Current Opinion in Environmental Sustainability*, 1, 170–178.

Hendrikx, J. and Hreinsson, E. Ö. (2010) The Potential Impact of Climate Change on Seasonal Snow Conditions in New Zealand. NIWA Client Report: CHC2010–153, November 2010.

Hennessy, K., Whetton, P. H., Walsh, K., Smith, I. N., Bathols, J. M., Hutchinson, M. and Sharples, J. (2008) Climate Change Effects on Snow Conditions in Mainland

Australia and Adaptation at Ski Resorts through Snow Making. *Climate Research*, 35, 255–270.

Higham, J. E. S. and Cohen, S. A. (2011) Canary in the Coalmine: Norwegian Attitudes Towards Climate Change and Extreme Long-haul Air Travel to Aotearoa/New Zealand. *Tourism Management*, 32(1), 98–105.

Hihara, K. (2010) Analysis on Bargaining about Global Climate Change Mitigation in International Aviation Sector. *Transportation Research Part E: Logistics and Transportation Review*, 47(3), 342–358.

Hirabayashi, Y., Kanae, S., Emori, S., Oki, T. and Kimoto, M. (2008) Global Projections of Changing Risks of Floods and Droughts in a Changing Climate. *Hydrological Sciences–Journal–des Sciences Hydrologiques*, 53(4), 754–772.

Hirsch, R. L. (2008) Mitigation of Maximum World Oil Production: Shortage Scenarios. *Energy Policy*, 36, 881–889.

HM Revenue and Customs (2010) Air Passenger Duty. Available at (12/12/2010) http://customs.hmrc.gov.uk/channelsPortalWebApp/channelsPortalWebApp.portal?_nfpb=trueand_pageLabel=pageExcise_ShowContentandpropertyType=documentandid=HMCE_CL_000505#P9_212

Hoerling, M., Eischeid, J., Perlwitz, J., Quan, X. W., Zhang, T. and Pegion, P. (2011) On the Increased Frequency of Mediterranean Drought. *Journal of Climate*, 24, 4519–4528.

Hoffman, A. (2011) Talking Past Each Other? Cultural Framing of Skeptical and Convinced Logics in the Climate Change Debate. *Organization and Environment*, 24(1), 3–33.

Hoffmann, V. H., Sprengel, D. C., Ziegler, A., Kolb, M. and Abegg, B. (2009) Determinants of Corporate Adaptation to Climate Change in Winter Tourism: An Econometric Analysis. *Global Environmental Change*, 19, 256–264.

Houston, J. R. (2008) The Economic Value of Beaches – A 2008 Update. *Shore and Beach*, 76(3), 22–26.

Howitt, O., Revol, V. G. N., Smith, I. J. and Rodger, C. J. (2010) Carbon Emissions from International Cruise Ship Passengers' Travel to and from New Zealand. *Energy Policy*, 38(5), 2552–2560.

Hoy, A., Hänsel, S. and Matschullat, J. (2011) How Can Winter Tourism Adapt to Climate Change in Saxony's Mountains? *Regional Environmental Change*, 11(3), 459–469.

Hudson, S. and Miller, G. A. (2005) Best Practices in Responsible Marketing of Tourism: The Case of Canadian Mountain Holidays. *Tourism Management*, 26(2), 133–142.

Huebner, A. (2011) Tourism and the (Un)Expected: A Research Note. *Pacific News*, 36, 25–28.

Hughes, L. and Rudolph, J. (2011) Future World Oil Production: Growth, Plateau, or Peak? *Current Opinion in Environmental Sustainability*, 3(4), 225–234.

Hunter, C. (1998) Sustainable Tourism as an Adaptive Paradigm. *Annals of Tourism Research*, 24(4), 850–867.

IATA (2008) Traffic Slowdown Continues – Asia Leads August Decline. Press Release. September 2008. Available at (10/10/08) www.iata.org/pressroom/pr/2008–09–30–01.html

IATA (2009a) A Global Approach to Reducing Aviation Emissions. Available at (02/01/2011) www.iata.org/SiteCollectionDocuments/Documents/Global_Approach_Reducing_Emissions_251109web.pdf

IATA (2009b) The Technology Roadmap Report. Available at (02/01/2011) www.iata.org/SiteCollectionDocuments/Documents/Technology_Roadmap_May2009.pdf

IATA (2011) Robust Recovery in 2010 Amid Changing Industry Trends. Available at (15/01/2011) www2.icao.int/en/NewsRoom/Lists/News/DispForm.aspx?ID=31

IFC (2009) Climate Risk and Business. Ports. Terminal Marítimo Muelles el Bosque, Cartagena, Colombia. International Finance Corporation, Washington, DC, 190pp.

Instinct and Reason (2008) Propensity for UK and German Travelers to Adapt Travel Intentions due to Rising Awareness of Climate Change Issues. Prepared for Department of Resources, Energy and Tourism, 2 May 2008. Surry Hills, NSW, Australia.

Institute for Integrative Tourism and Development (2009) Tourism Development in a Changing Climate. Background and Perspectives on the Role of Tourism in International Climate Politics. Available at (20/10/2010) www.respect.at

International Ecotourism Society, The (2009) Protecting Where the Wild Things Are. Available at (12/01/2011) www.yourtravelchoice.org/2009/04/protecting-where-the-wild-things-are/

International Energy Agency (2010) Energy Security. Available at (24/05/2011) www.iea.org/subjectqueries/keyresult.asp?KEYWORD_ID=4103

International Tourism Partnership, The (2010) Environmental Management for Hotels. Available for purchase at (08/02/2012) www.tourismpartnership.org/itp-shop/environmental-management-for-hotels-online-edition

International Union of Railways (2010) Keeping Climate Change Solutions on Track. The Role of Rail – A Global Position Paper. International Union of Railways, Paris, France, 44pp.

IPCC (2007a) Fourth Assessment Report (AR4). Summary for Policymakers. Synthesis Report. Contribution of Working Groups I, II and III to the Fourth Assessment Report of the Intergovernmental Panel on Climate Change. Geneva, Switzerland. Available at (20/03/2011) www.ipcc.ch/publications_and_data/publications_ipcc_fourth_assessment_report_synthesis_report.htm

IPCC (2007b) Climate Change 2007. Mitigation of Climate Change. Working Group II Contribution to the Fourth Assessment. Summary for Policymakers and Technical Summary.

IPCC (2011) Summary for Policy Makers, Special Report on Managing the Risks of Extreme Events and Disasters to Advance Climate Change Adaptation (SREX). Available at (20/11/2011) http://ipcc-wg2.gov/SREX

IPCC (no date) Emission Factor Database. Available at (06/01/2011) www.ipcc-nggip.iges.or.jp/EFDB/main.php

Jakobsson, K., Söderbergh, B., Höök, M. and Aleklett, K. (2009) How Reasonable are Oil Production Scenarios from Public Agencies? *Energy Policy*, 37(11), 4809–4818.

Jamaica Tourism Task Force (2009) Tourism Sector Plan, 2009–2030. Tourism Sector Plan for Vision 2030 Jamaica, Tourism Task Force, Kingston, Jamaica, 97pp.

Jariabka, M. (2008) Arctic Transform: Transatlantic Policy Options for Supporting Adaptation in the Marine Arctic. Arctic Transform Expert Workshop, Berlin, Germany, September 2008, Ecologic, Bonn, Germany, 6pp.

Jeuring, J. (2011) Intentions to Seek Severe Weather Information among Travelers in New Zealand. Master's Thesis – Tourism, Leisure and Environment, Wageningen University and Research Centre, Netherlands.

Jones, A. and Phillips, M. (2011) *Disappearing Destinations: Climate Change and the Future Challenges for Coastal Tourism*. Wallingford: CABI, 273pp.

Jones, B. and Scott, D. (2006) Climate Change, Seasonality and Visitation to Canada's National Parks. *Journal of Parks and Recreation Administration*, 24(2), 42–62.

Jones, S. (2011) The Impact of Climate Change on Tourism in the Rwenzori Mountains, Uganda, and Adaptation Strategies. Presented at Meeting the Challenges of Climate Change to Tourism in Africa and the Developing World. International Institute of Peace through Tourism, 15–20 May, Lusaka, Zambia.

Jopp, R., DeLacy, T., and Mair, J. (2010) Developing a Framework for Regional Destination Adaptation to Climate Change. *Current Issues in Tourism*, 13(6), 591–605.

Jorgensen, M. W. and Sorenson, S. C. (1997). Estimating emissions from railway traffic. Report for the project MEET: Methodologies for estimating air pollutant emissions from transport. Department of Energy Engineering, Technical University of Denmark - Available at (10/03/2005) www.inrets.fr/nois/infos/cost319/MEETDeliverables.pdf

Keesman, A. T., Blanco, J. M., Schillemans, R. A. A., Quiros, K. and Keskin, B. (2006) Drivers and Barriers for Energy Efficiency in the Hotel Sector in Costa Rica. Available at (20/05/2010) www.bun-ca.org/peest2/Arch_publi/ECLAT%20Final%20Version%20Word.pdf

Kelly, J. and Williams, P. (2007) Modelling Tourism Destination Energy Consumption and Greenhouse Gas Emissions: Whistler, British Columbia, Canada. *Journal of Sustainable Tourism*, 15(1), 67–90.

Klein, A., Kavoussi, S. and Lee, R. S. (2009) Weather Forecast Accuracy: Study of Impact on Airport Capacity and Estimation of Avoidable Costs. Eighth USA/Europe Air Traffic Management Research and Development Seminar, ATM2009, 10pp.

Kozak, N., Uysal, M. and Birkan, I. (2008) An Analysis of Cities based on Tourism Supply and Climatic Conditions in Turkey. *Tourism Geographies*, 10(1), 81–97.

Krarup, S. and Ramesohl, S. (2002) Voluntary Agreements on Energy Efficiency in Industry – Not a Golden Key, but Another Contribution to Improve Climate Policy Mixes. *Journal of Cleaner Production*, 10, 109–120.

Kuo, N. and Chen, P. (2009) Quantifying Energy Use, Carbon Dioxide Emission, and Other Environmental Loads from Island Tourism based on a Life Cycle Assessment Approach. *Journal of Cleaner Production*, 17, 1324–1330.

Kwon, T. H. (2005) A Scenario Analysis of CO_2 Emission Trends from Car Gravel: Great Britain 2000–2003. *Transport Policy*, 12(2), 175–184.

Lai, T. M., To, W. M., Lo, W. C., Choy, Y. S. and Lam, K. H. (2011) The Causal Relationship between Electricity Consumption and Economic Growth in a Gaming and Tourism Center: The Case of Macao SAR, the People's Republic of China. *Energy*, 36(2), 1134–1142.

Lambert, E., Hunter, C., Pierce, G. J. and MacLeod, C. (2010) Sustainable Whale-watching Tourism and Climate Change towards a Framework of Resilience. Special Issue (edited by Scott, D. and Becken, S.): Tourism: Adapting to Climate Change and Climate Policy. *Journal of Sustainable Tourism*, 18(3), 409–428.

Lassen, C. (2010) Environmentalist in Business Class: An Analysis of Air Travel and Environmental Attitudes. *Transport Reviews*, 30(6), 733–751.

Lee, D. S., Fahey, D. W., Forster, P. M., Newton, P. J., Wit, R. C. N., Lim, L. L., Owen, B. and Sausen, R. (2009) Aviation and Global Climate Change in the 21st century. *Atmospheric Environment*, 43(22–23), 3520–3537.

Lee, J. J. (2010) Can we Accelerate the Improvement of Energy Efficiency in Aircraft Systems? *Energy Conversion and Management*, 51, 189–196.

Lee, S. and Park, S. (2010) Financial Impacts of Socially Responsible Activities on Airline Companies. *Journal of Hospitality and Tourism Research*, 34(2), 185–203.

Lehmann, L. V. (2009) The Relationship between Tourism and Water in Dry Land Regions. Presented at the Environmental Research Event 2009, Noosa, QLD. Available at (20/11/2011) http://espace.library.uq.edu.au/eserv/UQ:179617/Lehmann_-_ERE2009.pdf

Leigh, C. (2011) New Tourism in a New Society Arises from 'Peak Oil'. *Tourismos*, 6(1). Available at (20/09/2011) www.chios.aegean.gr/tourism/journal.htm

Lemelin, H., Dawson, J., Stewart, E. J., Maher, P. and Lueck, M. (2010) Lastchance Tourism: The Boom, Doom, and Gloom of Visiting Vanishing Destinations, *Current Issues in Tourism*, 13(5), 477–493.

Lin, T. (2010) Carbon Dioxide Emissions from Transport in Taiwan's National Parks. *Tourism Management*, 31(2), 285–290.

Linham, M. and Nicholls, R. J. (2010) Technologies for Climate Change Adaptation – Coastal Erosion and Flooding. UNEP Risø Centre on Energy, Climate and Sustainable Development, Risø DTU National Laboratory for Sustainable Energy, Roskilde, Denmark, 166pp.

Lobo, A. (2007) Tourism and Climate Change: How can Portuguese Tourism Sector Contribute to Reduce Energy Consumption and Pressure on CO_2 Emissions? Prospectiva e Planeamento, 14. Available at (12/12/2010) www.dpp.pt/pages/files/Turismo.pdf

Lonely Planet (2007) Annual Travellers Pulse Survey. Available at (20/05/08) www.lonelyplanet.com/pressroom/news/press_release.cfm?press_release_id=313

Lucas, C., Hennessy, K., Mills, G., and Bathols, J. (2007) *Bushfire Weather in Southeast Australia: Recent Trends and Projected Climate Change Impacts*. Melbourne: Bushfire Cooperative Research Centre, 57pp.

Lynes, J. and Andrachuk, M. (2008) Motivations for Corporate Social and Environmental Responsibility: A Case Study of Scandinavian Airlines. *Journal of International Management*, 14(4), 377–390.

Macintosh, A. and Wallace, L. (2009) International Aviation Emissions to 2025: Can Emissions be Stabilized without Restricting Demand? *Energy Policy*, 37, 264–273.

Maibach, M. and Schneider, C. (2002) External Costs of Corridors. A Comparison between Air, Road and Rail. Air Transport Action Group. Final Report, Zurich. Available at (20/10/04) www.atag.org/content/showpublications.asp?folderid=435andlevel1=4andlevel2=435and

Mair, J. (2011) Exploring Air Travellers' Voluntary Carbon-offsetting. *Journal of Sustainable Tourism*, 19(2), 215–230.

Management Consultancies Association (2007) A Growing Concern: How Should Business Adapt to a Low carbon World? Available at (09/12/2009) www.mca.org.uk

Marshall, N., Marshall, P., Abdulla, A., Rouphael, T. and Amr, A. (2011) Preparing for Climate Change: Recognising its Early Impacts through the Perceptions of Dive Tourists and Dive Operators in the Egyptian Red Sea. *Current Issues in Tourism*, 14(6), 507–518.

Martín-Cejas, R. R. and Sánchez, P. P. R. (2010) Ecological Footprint Analysis of Road Transport related to Tourism Activity: The Case for Lanzarote Island. *Tourism Management*, 31(1), 98–103.

Matasci, C. and Altamirano-Cabrera, J. C. (2010) Climate Change and Tourism in Switzerland – a Survey on Impacts, Vulnerability and Possible Adaptation Measures. Research Paper 2010/05, NCCR Climate, Swiss Climate Research and University of Bern, Switzerland, 23pp.

Matson, J. (2010) What will Space Tourism mean for Climate Change? *Scientific American*, Oct 23. Available at (12/11/2010) www.scientificamerican.com/blog/post.cfm?id=what-will-space-tourism-mean-for-cl-2010–10–23

Mayes, G., Smith, T. and Thomsen, D. (2010) Unique Responses by the Sheraton Resort Fiji to the Impacts of Climate Change. Paper presented at the 2010 International Climate Change Adaptation Conference, Climate Adaptation Futures, Gold Coast. Available at (08/02/2012) www.nccarf.edu.au/conference2010/archives/1071

Mayor, K. and Tol, R. (2007) The Impact of the UK Aviation Tax on Carbon Dioxide Emissions and Visitor Numbers. *Transport Policy*, 14, 507–513.

Mayor, K. and Tol, R. (2008) The Impact of the EU-US Open Skies Agreement on International Travel and Carbon Dioxide Emissions. *Journal of Air Transport Management*, 14, 1–7.

Mayor, K. and Tol, R. (2010) Scenarios of Carbon Dioxide Emissions from Aviation. *Global Environmental Change*, 20, 65–73.

McDevitt, A. (2010) Topic Guide: Climate Change and Adaptation. Governance and Social Development Resource Center (GSDRC), International Development Department, College of Social Sciences, University of Birmingham, UK, 43pp.

McEvoy, D., Handley, J. F., Cavan, G., Aylen, J., Lindley, S., McMorrow, J. and Glynn, S. (2006) Climate Change and the Visitor Economy. Challenges and Opportunities for England's Northwest. Sustainability Northwest and UKCIP, Oxford, 40pp.

McGray, H., Hammill, A. and Bradley, R. (2007) Weathering the Storm: Options for Framing Adaptation and Development, World Resources Institute, Washington, DC. Available at (01/04/2011) http://pdf.wri.org/weathering_the_storm.pdf

McKercher, B., Prideaux, B., Cheung, C. and Law, R. (2010) Achieving Voluntary Reductions in the Carbon Footprint of Tourism and Climate Change. Special Issue (edited by Scott, D. and Becken, S.): Tourism: Adapting to Climate Change and Climate Policy. *Journal of Sustainable Tourism*, 18(3), 297–318.

McMullen, C. P. and Jabbour, J. (2009) Climate Change Science Compendium 2009. United Nations Environment Programme, Nairobi, EarthPrint, 76pp.

Mensah, I. (2006) Environmental Management Practices among Hotels in the Greater Accra Region. *International Journal of Hospitality Management*, 25, 414–431.

Met Office (2010) Reliability: Keeping you Moving with our Weather and Climate Change Services for Rail. Met Office, Exeter, United Kingdom, 4pp.

Michalena, E., Hills, J. and Amat, J. P. (2009) Developing Sustainable Tourism, Using a Multicriteria Analysis on Renewable Energy in Mediterranean Islands. *Energy for Sustainable Development*, 13, 129–136.

Moreno, A. (2010) Mediterranean Tourism and Climate (Change): A Survey-based Study. *Tourism and Hospitality Planning and Development*, 7(3), 253–265.

Moreno, A., Amelung, B. and Santamarta, L. (2008) Linking Beach Recreation to Weather Conditions. A Case Study in Zandvoort, Netherlands. *Tourism in Marine Environments*, 5(2/3), 111–120.

Moreno, A. and Becken, S. (2009) A Climate Change Vulnerability Assessment Methodology for Coastal Tourism. *Journal of Sustainable Tourism*, 17(4), 473–488.

Moriarty, P. and Honnery, D. (2008a) Low-mobility: The Future of Transport. *Futures*, 40(10), 865–872.

Moriarty, P. and Honnery, D. (2008b) The Prospects for Global Green Car Mobility. *Journal of Cleaner Production*, 16, 1717–1726.

Morrell, P. (2009) The Potential for European Aviation CO_2 Emissions Reduction through the Use of Larger Jet Aircraft. *Journal of Air Transport Management*, 15, 151–157.

Müller, H. and Weber, F. (2008) Climate Change and Tourism – Scenario Analysis for the Bernese Oberland in 2030. *Tourism Review*, 63(3), 57–71.

NASA (2010) FACET: Future Air Traffic Management Concepts Evaluation Tool. NASA Facts, Ames Research Center, Moffett Field, California, National Aeronautics and Space Administration (NASA), 2pp.

Needham, M. (2010) Value Orientations Toward Coral Reefs in Recreation and Tourism Settings: A Conceptual and Measurement Approach. *Journal of Sustainable Tourism*, 18(6), 735–756.

Nelson, V. (2010) Promoting Energy Strategies on Eco-Certified Accommodation Websites. *Journal of Ecotourism*, 9(3), 187–200.

Nepal, S. K. (2008) Tourism-induced Rural Energy Consumption in the Annapurna Region of Nepal. *Tourism Management*, 29, 89–100.

Network Rail (2010) Interim Climate Change Adaptation Report, Network Rail, London, United Kingdom, 22pp.

New Zealand Ministry of Tourism (2008) New Zealand Tourism and Climate Change Plan. Available at (11/09/2009) http://www.tourism.govt.nz/Our-Work/Our-Work Summary-page/Climate-Change/

Nicholls, S. and Holecek, D. (2008) Engaging Stakeholders in the Development of Climate Change Decision-support Tools: A Case Study from Michigan, USA. *Tourism Review International*, 12, 25–42.

Nickson, A., Woolston, H., Daniels, J., Dedring, I., Reid, K., Ranger, K., Clancy, L., Street, R., and Reeder, T. (2011) Managing Risks and Increasing Resilience. The Mayor's Climate Change Adaptation Strategy. Greater London Authority, London, United Kingdom, 126pp.

Nicolau, J. L. and Mas, F. J. (2006) The Influence of Distance and Prices on the Choice of Tourist Destinations: The Moderating Role of Motivations. *Tourism Management*, 27, 982–996.

NOAA (2011) 2010 Tied for Warmest Year on Record. Available at (28/09/2011) www.noaanews.noaa.gov/stories2011/20110112_globalstats.html

Nolte, R. (2010) ARISCC – Adaptation of Railway Infrastructure to Climate Change. Workshop Presentation, October 2010, Paris.

Nygren, E., Aleklett, K. and Höök, M. (2009) Aviation Fuel and Future Oil Production Scenarios. *Energy Policy*, 37, 4003–4010.

OECD (2001) Environmental Outlook. By Organisation for Economic Co-operation and Development. Paris: OECD Publishing.

OECD (2005) Bridge Over Troubled Waters: Linking Climate Change and Development. Organisation for Economic Cooperation and Development (OECD), Paris, France, 157pp.

OECD (2009) OECD Policy Guidance on Integrating Climate Change Adaptation into Development Co-operation. Organization for Economic Co-operation and Development (OECD), Paris, France, 197pp.

OECD and UNEP (2011) Climate Change and Tourism Policy in OECD Countries. Organisation for Economic Co-operation and Development (OECD) and the United Nations Environment Programme (UNEP), Paris, France, 100pp.

Orchiston, C. (2010) Tourism and Seismic Risk: Perceptions, Preparedness and Resilience in the Zone of the Alpine Fault, Southern Alps, New Zealand. A thesis submitted for the degree of Doctor of Philosophy at the University of Otago Dunedin, New Zealand. March 2010.

Otto, I. and Heath, E. T. (2009) The Potential Contribution of the 2010 Soccer World Cup to Climate Change: An Exploratory Study among Tourism Industry Stakeholders in the Tshwane Metropole of South Africa. *Journal of Sport and Tourism*, 14(2–3), 169–191.

Overstrand Municipality (2010) Annual Report 2009/2010, Overstrand Municipality, South Africa, 298pp.

Parkman, M. (2008) Assessing the Impact of Climate Change on Transport Infrastructure. Rail Safety and Standards Board Ltd., London, United Kingdom, 110pp.

Parliamentary Office of Science and Technology (2010) Resilience of UK Infrastructure. POSTNOTE Number 362, October 2010, Parliamentary Office of Science and Technology, London, United Kingdom, 4pp.

Pearce, T., Ford, J., Duerden, F., Smit, F., Andrachuk, M., Berrang-Ford, L. and Smith, T. (2010) Advancing Adaptation Planning for Climate Change in the Inuvialuit Settlement Region (ISR): A Review and Critique. *Regional Environmental Change*. Available at (08/02/2012) www.cakex.org/sites/default/files/Pearce%20et%20al.%202010%20REC.pdf

Peeters, P. (2010) Tourism Transport, Technology, and Carbon Dioxide Emissions. Chapter 4 (pp. 67–90). In Schott, C. (ed.), *Tourism and the Implications of Climate Change: Issues and Actions*. Bridging Tourism Theory and Practice. Volume 3. Bingley: Emerald.

Peeters, P. and Middel, J. (2007) Historical and Future Development of Air Transport Fuel Efficiency. In Sausen *et al.* (Eds.). Proceedings of the International Conference on Transport, Atmosphere and Climate, Oxford, 26–29 June 2006.

Peeters, P., Middel, J. and Hoolhorst A. (2005) Fuel Efficiency of Commercial Aircraft: An Overview of Historical and Future Trends. Available at (08/02/2012) www.transportenvironment.org/search/site/fuel%20efficiency%20of%20commercial%20aircraft

Peeters, P., Szimba, E. and Duijnisveld, M. (2007) Major Environmental Impacts of European Tourist Transport. *Journal of Transport Geography*, 15, 83–93.

Peeters, P., van Egmond, T. and Visser, N. (2004) *European Tourism, Transport and Environment*. Final Version. Breda: NHTV CSTT.

Pejovic, T. (2008) Implications of Climate Change for the UK Aviation Sector. A thesis for the degree of Doctor of Philosophy, Centre for Transport Studies, Department of Civil and Environmental Engineering, Imperial College, London, United Kingdom, 220pp.

Pejovic, T., Williams, V. A., Noland, R. B. and Toumi, R. (2009) Factors Affecting the Frequency and Severity of Airport Weather Delays and the Implications of Climate Change for Future Delays. *Journal of the Transportation Research Board*, 2139, 97–106.

Peng, L. and Guihua, Y. (2007) Ecological Footprint Study on Tourism Itinerary Products in Shangri-La, Yunnan Province, China. *Acta Ecologica Sinica*, 27(7), 2954–2963.

Penner, J., Lister, D., Griggs, D., Dokken, D. and McFarland, M. (Eds.) (1999) *Aviation and the Global Atmosphere*. A Special Report of IPCC Working Groups I and III. Published for the Intergovernmental Panel on Climate Change. Cambridge: Cambridge University Press.

Pentelow, L. and Scott, D. (2011) Aviation's Inclusion in International Climate Policy Regimes: Implications for the Caribbean Tourism Industry. *Journal of Air Transport Management*, 17, 199–205.

Perch-Nielsen, S. (2009) The Vulnerability of Beach Tourism to Climate Change: An Index Approach. *Climatic Change*, 100(3–4), 769–815.

Perch-Nielsen, S., Sesartic, A. and Stucki, M. (2010) The Greenhouse Gas Intensity of the Tourism Sector: The Case of Switzerland. *Environmental Science and Policy*, 13, 131–140.

Persad, J. (2011) Regional Monitoring and Evaluation System for Disaster Risk Management and Climate Change Adaptation in The Caribbean Tourism Sector. Caribbean Disaster Emergency Management Agency (CDEMA), Barbados, West Indies, 21pp.

Peterson, T. C., McGuirk, M., Houston, T. G., Horvitz, A. H. and Wehner, M. S. (2008) Climate Variability and Change with Implications for Transportation. Commissioned Paper for Potential Impacts of Climate Change on U.S. Transportation. Special Report 290, Transportation Research Board, Washington, DC, 90pp.

Pew (2009) Public Praises Science; Scientists Fault Public, Media. News Release, Pew Research Center for the People and the Press, Washington, DC, 100pp.

Pham, T., Simmons, D. and Spurr, R. (2010) Climate Change-induced Economic Impacts on Tourism Destinations: The Case of Australia. Special Issue (edited by Scott, D. and Becken, S.): Tourism: Adapting to Climate Change and Climate Policy. *Journal of Sustainable Tourism*, 18(3), 449–473.

Pickering, C. M. and Buckley R. C. (2010) Climate Response by the Ski Industry: The Shortcomings of Snowmaking for Australian Resorts. *AMBIO: A Journal of the Human Environment*, 39(5–6): 430–438.

Pickering, C. M., Castley, J. G. and Burtt, M. (2009) Skiing Less Often in a Warmer World: Attitudes of Tourists to Climate Change in an Australian Ski Resort. *Geographical Research*, 48(2), 137–147.

Prideaux, B. (2000) The Resort Development Spectrum: A New Approach to Modeling Resort Development. *Tourism Management*, 21(3), 225–240.

Prideaux, B. (2004) The Resort Development Spectrum: The Case ofthe Gold Coast, Australia. *Tourism Geographies*, 6(1), 26–58.

Prideaux, B., Coghlan, A. and McNamara, K. (2010) Assessing Tourists' Perceptions of Climate Change on Mountain Landscapes. *Tourism Recreation Research*, 35(2), 187–199.

Priyadarsini, R., Xuchao, W. and Eang, L. S. (2009) A Study on Energy Performance of Hotel Buildings in Singapore. *Energy and Buildings*, 41, 1319–1324.

Prowse, M. and Snilstveit, B. (2009) Impact Evaluation and Interventions to Address Climate Change: A Scoping Study. International Institute for Impact Evaluation, Global Development Network, New Delhi, India, 66pp.

Rail Safety and Standards Board (2008) Assessing the Impact of Climate Change on Transport Infrastructure. Rail Safety and Standards Board Ltd., London, United Kingdom, 110pp.

Rail Safety and Standards Board (2010) Tomorrow's Railway and Climate Change Adaptation. Phase 1 Report, Rail Safety and Standards Board Ltd., London, United Kingdom, 20pp.

Rendeiro Martín-Cejas, R. and Ramírez Sánchez, P. (2010) Ecological Footprint Analysis of Road Transport related to Tourism Activity: The Case for Lanzarote Island. *Tourism Management*, 31(1), 98–103.

Ribeiro, M. M., Losenno, C., Dworak, T., Massey, E., Swart, R., Benzie, M. and Laaser, C. (2009) Design of Guidelines for the Elaboration of Regional Climate Change Adaptations Strategies. Study for European Commission – DG Environment. Ecologic Institute, Vienna, Austria, 91pp.

Richardson, K. *et al.* (2011) Synthesis Report from Climate Change: Global Risks, Challenges and Decisions Conference, Copenhagen, 10–12 March 2009, University of Copenhagen, Denmark, 39pp.

Ritchie, B. W. (2004) Chaos, Crises and Disasters: A Strategic Approach to Crisis Management in the Tourism Industry. *Tourism Management*, 25(6), 669–683.

Robertson, D., Kean, I. and Moore, S. (2006) Tourism Risk Management: An Authoritative Guide to Managing Crises in Tourism. APEC International Centre for Sustainable Tourism (AICST), Asia-Pacific Economic Cooperation (APEC) Secretariat, Singapore, 98pp.

Rossello-Batle, B., Moia, A., Cladera, A. and Martinez, V. (2010) Energy Use, CO_2 Emissions and Waste throughout the Life Cycle of a Sample of Hotels in the Balearic Islands. *Energy and Buildings*, 42, 547–558.

Rosselló-Nadal, J., Riera-Font, A. and Cardenas, V. (2011) The Impact of Weather Variability on British Outbound Flows. *Climatic Change*, 105, 281–292.

Rossetti, M. (2002) Potential Impacts of Climate Change on Railroads. In The Potential Impacts of Climate Change on Transportation Workshop, October 2002, Center for Climate Change and Environmental Forecasting, U.S. Department of Transportation, Washington, DC, 1–13.

Royal Academy of Engineering (2011) Infrastructure, Engineering and Climate Change Adaptation: Ensuring Services in an Uncertain Future. The Royal Academy of Engineering, London, United Kingdom, 108pp.

Rummukainen, M., Räisänen, J., Björnsson, H. and Christensen, J. H. (2010) Physical Climate Science since IPCC AR4. A brief update on new findings between 2007 and April 2010. Nordic Council of Ministers, Copenhagen, Denmark, 91pp.

Rutty, M. and Scott, D. (2010) Will the Mediterranean Become 'Too Hot' for Tourism? A Reassessment. *Tourism and Hospitality Planning and Development*, 7(3), 267–281.

Samoa Tourism Authority (2009) Samoa Tourism Development Plan (2009–2013). Samoa Tourism Authority, Apia, Samoa, 143pp.

Sanahuja, H. E. (2011) Tracking Progress for Effective Action: A Framework for Monitoring and Evaluating Adaptation to Climate Change. Community of Practice, Global Environment Facility, Washington, DC, 78pp.

Sanders, D., Laing, J. and Houghton, M. (2008) Tourism Recovery: The Impact of Bushfires on Tourism and Visitation in Alpine National Parks. Sustainable Tourism CRC, Australia, 16pp.

Schernewski, G. (2009) Climate Change in the Baltic: An Opportunity for Tourism and Rural Development? *Coastal and Marine*, 18(2), 12.

Schiermeier, Q. (2010) The Real Holes in Climate Science. *Nature*, 463, 284–287.

Schiff, A. and Becken, S. (2011) Demand Elasticities for Tourism in New Zealand. *Tourism Management*, 32(3), 564–575.

Scott, D. and Becken, S. (2010) Adapting to Climate Change and Climate Policy: Progress, Problems and Potentials. Special Issue (edited by Scott, D. and Becken, S.): Tourism: Adapting to Climate Change and Climate Policy. *Journal of Sustainable Tourism*, 18(3), 283–295.

Scott, D., Dawson, J. and Jones, B. (2011) Climate Change Vulnerability of the US Northeast Winter Recreation – Tourism Sector. *Mitigation and Adaptation Strategies for Global Change*, 13(5–6), 577–596.

Scott, D., Gössling, S. and de Freitas, C. (2008) Preferred Climate for Tourism: Case Studies from Canada, New Zealand and Sweden. *Climate Research*, 38, 61–73.

Scott, D. and Jones, B. (2007) A Regional Comparison of the Implications of Climate Change of the Golf Industry in Canada. *The Canadian Geographer*, 51(2), 219–232.

Scott, D., Jones, B. and Konopek, J. (2007) Implications of Climate and Environmental Change for Nature-based Tourism in the Canadian Rocky Mountains: A Case Study of Waterton Lakes National Park. *Tourism Management*, 28(2), 570–579.

Scott, D. and Lemieux, C. (2009) Weather and Climate Information for Tourism. White Paper, commissioned by the World Meteorological Organisation.

Scott, D. and McBoyle, G. (2007) Climate Change Adaptation in the Ski Industry. *Mitigation and Adaptation Strategies for Global Change*, 12(8), 1411–1431.

Scott, D., McBoyle, G., Mills, B. and Minogue, A. (2006) Climate Change and the Sustainability of Ski-based Tourism in Eastern North America. *Journal of Sustainable Tourism*, 14(4), 376–398.

Scott, D., Peeters, P. and Gössling, S. (2010) Can Tourism Deliver its 'Aspirational' Greenhouse Gas Emission Reduction Targets? Special Issue (edited by Scott, D. and Becken, S.): Tourism: Adapting to Climate Change and Climate Policy. *Journal of Sustainable Tourism*, 18(3), 393–408.

Serquet, G. and Rebetez, M. (2011) Relationship Between Tourism Demand in the Swiss Alps and Hot Summer Air Temperatures associated with Climate Change. *Climatic Change*, 108(1–2), 291–300.

Sgouridis, S., Bonnefoy, P. A. and Hansman, R. J. (2010) Air Transportation in a Carbon Constrained World: Long-term Dynamics of Policies and Strategies for Mitigating the Carbon Footprint of Commercial Aviation. *Transportation Research Part A: Policy and Practice*, 45(10), 1077–1091.

Shih, C., Nicholls, S. and Holecek, D. (2009) Impact of Weather on Downhill Ski Lift Ticket Sales. *Journal of Travel Research*, 47(3), 359–372.

Simpson, M. C., Gössling, S. and Scott, D. (2008a) Report on the International Policy and Market Response to Global Warming and the Challenges and Opportunities that Climate Change Issues Present for the Caribbean Tourism Sector. Caribbean Regional Sustainable Tourism Development Programme (CRSTDP), Caribbean Tourism Organization, Barbados.

Simpson, M. C., Gössling, S., Scott, D., Hall, C. M. and Gladin, E. (2008b) Climate Change Adaptation and Mitigation in the Tourism Sector: Frameworks, Tools and Practices. UNEP, University of Oxford, UNWTO, WMO, Paris, France, 152pp.

Simpson, M., Scott, D. and Trotz, U. (2011) Climate Change's Impact on the Caribbean's Ability to Sustain Tourism, Natural Assets, and Livelihoods. Technical Note No. IDB-TN-238, Environmental Safeguards Unit, Inter-American Development Bank (IDB), Washington, DC, 19pp.

Sisman and Associates (2007) Tourism Destinations Carbon Footprints. Available at (10/04/2011) www.teamfortheworld.org/docs/articles/tourismdestinations.pdf

Smerecnik, K. R. and Andersen, P. A. (2011) The Diffusion of Environmental Sustainability Innovations in North American Hotels and Ski Resorts. *Journal of Sustainable Tourism*, 19(2), 171–197.

SNV Netherlands Development Organisation (2009) The Market for Responsible Tourism Products. SNV Netherlands Development Organisation, Asia Regional Office, Hanoi, Vietnam, 178pp.

Sorrell, S., Speirs, J., Bentley, R., Brandt, A. and Miller, R. (2010) Global Oil Depletion: A Review of the Evidence. *Energy Policy*, 38, 5290–5295.

Statistics Greenland (2011) Statistical Yearbook. Available at (05/02/2012) www.stat.gl/dialog/main.asp?lang=en&version=2011&link=SA&colcode=a&version=2011

Steele, P. (2009) Why We Need Alternative Fuels. IATA presentation. Available at (02/02/2011) www.icao.int/WAAF2009/Presentations/3_Steele.pdf

Stewart, E. J., Howell, S. E. L., Draper, D., Yackel, J. and Tivy, A. (2007) Sea Ice in Canada's Arctic: Implications for Cruise Tourism. *Arctic*, 60(4), 370–380.

Stewart, E. J., Tivy, A., Howell, S. E. L., Dawson, J. and Draper, D. (2010) Cruise Tourism and Sea Ice in Canada's Hudson Bay Region. *Arctic*, 63(1), 57–66.

Sunshine Coast Council (2010a) Climate Change and Peak Oil Strategy. Available at: (24/11/2010) www.sunshinecoast.qld.gov.au/sitePage.cfm?code=cc-strategy

Sunshine Coast Council (2010b) Climate Change Background Study, Climate Change and Peak Oil Strategy, 2010–2020. Sunshine Coast Council, Sunshine Coast, Australia, 88pp.

Sustainable Economic Development Unit (SEDU), Department of Economics (2002) Environmental Management Insertion in Tourism Sector Policies in the Caribbean. Final Report. University of the West Indies, St. Augustine. Available at (02/04/2011) http://idbdocs.iadb.org/wsdocs/getdocument.aspx?docnum=649465

Swart, R., Biesbroek, R., Binnerup, S., Carter, T. R., Cowan, C., Henrichs, T., Loquen, S., Mela, H., Morecroft, M., Reese, M. and Rey, D. (2009) Europe Adapts to Climate Change: Comparing National Adaptation Strategies. PEER Report No. 1, Partnership for European Environmental Research, Helsinki, Finland, 283pp.

Tervo, K. (2008) The Operational and Regional Vulnerability of Winter Tourism to Climate Variability and Change: The Case of the Finnish Nature-based Tourism Entrepreneurs. *Scandinavian Journal of Hospitality Research*, 8(4), 317–332.

Tirado, M. C., Clarke, R., Jaykus, L. A, McQuatters-Gollop, A. and Frank, J. M. (2010) Climate Change and Food Safety: A Review. *Food Research International*, 43, 1745–1765.

Tortella, D. and Tirado, D. (2011) Hotel Water Consumption at a Seasonal Mass Tourist Destination: The Case of the Island of Mallorca. *Journal of Environmental Management*, 92(10), 2568–2579.

Tourism Australia (2009) Tourism Australia's Brand Health Monitor. Attitudes to Climate Change and Environmental Degradation. March 2009. Available at (08/02/2012) www.ret.gov.au/tourism/Documents/Tourism%20Policiy/workshop%20presentations/090622%20Climate%20Change%20for%20Michelle%20Gapes.pdf

Tourism New Zealand (2008) Environment: Attitudes and Behaviour. Key Messages from the Campaign Tracking Research 06/07. Auckland, New Zealand.

Tourism Queensland (2007) Regional Tourism Crisis Management Plan Template. A Guide to Preparing a Regional Tourism Crisis Management Plan. Brisbane, Australia, 32pp.

Tourism Research Australia (2007) A National Framework for Best Practice Destination Management Planning. Tourism Research Australia, Department of Resources, Energy and Tourism, Government of Australia, Canberra, Australia, 10pp.

Tourism Victoria (2010) Crisis Essentials. Crisis Management for Tourism Businesses. Available at (05/12/2010) www.vic.gov.au/crisis

Transportation Research Board (2008) Potential Impacts of Climate Change on U.S. Transportation. Special Report 290, Transportation Research Board, Washington, DC, 298pp.

Tratalos, J. A., Gill, J. A., Jones, A., Showler, D., Bateman, I., Watkinson, A., Sugden, R. and Sutherland, W. J. (2005) Interactions between Tourism, Breeding Birds and Climate Change across a Regional Scale. Tyndall Centre for Climate Change Research, Technical Report 36.

Travel Foundation, The (2006) Insider Guide: Climate Change and Tourism: A Guide for Managers. November 2006, Bristol, UK. Available at (20/01/2011) www.thetravelfoundation.org.uk

Trung, D. N. and Kumar, S. (2004) Resource Use and Waste Management in Vietnam Hotel Industry. *Journal of Cleaner Production*, 13, 109–116.

Turton, S., Dickson, T., Hadwen, W., Jorgensen, B., Pham, T., Simmons, D., Tremblay, P. and Wilson, R. (2010) Developing an Approach for Tourism Climate Change Assessment: Evidence from Four Contrasting Australian Case Studies. Special Issue (edited by Scott, D. and Becken, S.): Tourism: Adapting to Climate Change and Climate Policy. *Journal of Sustainable Tourism*, 18(3), 429–448.

Turton, S., Hadwen, W., Wilson, R., Jorgensen, B. and Simmons, D. (2009) The Impacts of Climate Change on Australian Tourist Destinations. Developing Adaptation and Response Strategies. Sustainable Tourism Cooperative Research Centre. Gold Coast, Australia.

Uemura, Y., Kai, T., Natori, R., Takahashi, T., Hatate, Y. and Yoshida, M. (2003) Potential of Renewable Energy Sources and its Applications in Yakushima Island. *Renewable Energy*, 29, 581–591.

Uhlmann, B., Goyette, S. and Beniston, M. (2009) Sensitivity Analysis of Snow Patterns in Swiss Ski Resorts to Shifts in Temperature, Precipitation and Humidity under Conditions of Climate Change. *International Journal of Climatology*, 29(8), 1048–1055.

UK Energy Research Institute (2009) Global Oil Depletion: An Assessment of the Evidence for a Near-term Peak in Global Oil Production. London, United Kingdom.

UNCTAD (2009) Maritime Transport and the Climate Change Challenge. Summary of Proceedings, Multi-Year Expert Meeting on Transport and Trade Facilitation. United Nations Conference on Trade and Development (UNCTAD), UNCTAD/DTL/TLB/2009/1, Geneva, Switzerland, 52pp.

UNEP (2009) Sustainable Coastal Tourism: An Integrated Planning and Management Approach. Sustainable Consumption and Production Branch, United Nations Environment Programme (UNEP), Paris, France, 87pp.

UNEP and CAST (2008) Disaster Risk Management for Coastal Tourism Destinations. Responding to Climate Change. A Practical Guide for Decision Makers. UNEP DTIE: Paris. Available at (22/05/2010) www.unep.fr/scp

UNESCO (2007) Case Studies on Climate Change and World Heritage. UNESCO World Heritage Centre, Paris, France, 55pp.

UNESCO (2010) Managing Disaster Risks for World Heritage. United Nations Educational, Scientific and Cultural Organization (UNESCO), Paris, France, 69pp.

UNFCCC (2010) Synthesis Report on Efforts Undertaken to Monitor and Evaluate the Implementation of Adaptation Projects, Policies and Programmes and the Costs and Effectiveness of Completed Projects, Policies and Programmes, and Views on Lessons Learned, Good Practices, Gaps and Needs. United Nations Framework Convention on Climate Change (UNFCCC), Subsidiary Body for Scientific and Technological Advice, Thirty-second Session, Bonn, Germany, 31 May to 9 June 2010, Distr. GENERAL, FCCC/SBSTA/2010/5, 15pp.

UNFCCC (2011) The United Nations Climate Change Conference in Cancun, COP 16/ CMP 6, 29 November to 10 December 2010. Available at (02/04/2011) http://unfccc.int/meetings/cop_16/items/5571.php

UNISDR (2010) Local Governments and Disaster Risk Reduction: Good Practices and Lessons Learned. United Nations Secretariat of the International Strategy for Disaster Reduction (UNISDR), Geneva, Switzerland, 86pp.

UNWTO (2004) Guidebook on Indicators for Sustainability for Tourism Destinations. United Nations World Tourism Organisation (UNWTO), Madrid, Spain.

UNWTO (2007) Djerba Declaration on Tourism and Climate Change. Available at (02/08/2010) http://sdt.unwto.org/sites/all/files/docpdf/djerbadeclaration-eng.pdf

UNWTO (2010) Tourism Highlights, 2010 Edition. United Nations World Tourism Organisation (UNWTO), Madrid, Spain, 12pp.

UNWTO (2011) *UNWTO World Tourism Barometer*, 9(1), February 2011, Madrid, Spain.

UNWTO, UNEP and WMO (2008) *Climate Change and Tourism: Responding to Global Challenges*. Madrid: United Nations World Tourism Organization; Paris: United Nations Environment Program; Geneva: World Meteorological Organization.

Urban, F. and Mitchell, T. (2011) Climate Change, Disasters and Electricity Generation. Strengthening Climate Resilience Discussion Paper 8, Strengthening Climate Resilience (SCR), coordinated by the UK Institute of Development Studies, Plan International and Christian Aid, London, United Kingdom, 35pp.

Urry, J. (2008) Climate Change, Travel and Complex Futures. *The British Journal of Sociology*, 59(1), 261–279.

URS (2010) Adapting Energy, Transport and Water Infrastructure to the Long-term Impacts of Climate Change. Report Ref. No RMP/5456, prepared for the Government of the United Kingdom by URS Corporation Ltd., London, United Kingdom, 194pp.

USAID (2006) The Kathmandu Electric Vehicle Alliance (KEVA). Final Program Performance Report. Available at (18/01/2011) http://pdf.usaid.gov/pdf_docs/PDACI197.pdf

USEPA (2008) 2008 Planning for Climate Change Impacts at U.S. Ports. White Paper, United States Environmental Protection Agency (USEPA), Fairfax, US, 16pp.

USFDA (2009) Food Code 2009. US Department of Health and Human Services, Public Health Service, US Food and Drug Administration (USFDA), College Park, US, 698pp. Available at (05/02/2012) www.fda.gov/downloads/Food/FoodSafety/Retail FoodProtection/FoodCode/FoodCode2009/UCM189448.pdf

Uyarra M., Cote I., Gill, J., Tinch, R., Viner, D. and Watkinson, A. (2005) Island-specific Preferences of Tourists for Environmental Features: Implications of Climate Change for Tourism-dependent States. *Environmental Conservation*, 3(1), 11–19.

Verbruggen, A. and Al Marchohi, M. (2010) Views on Peak Oil and its Relation to Climate Change Policy. *Energy Policy*, 38(10), 5572–5581.

Villanueva, P. S. (2011) Learning to ADAPT: Monitoring and Evaluation Approaches in Climate Change Adaptation and Disaster Risk Reduction – Challenges, Gaps and Ways Forward. SCR Discussion Paper 9, Strengthening Climate Resilience (SCR), coordinated by the UK Institute of Development Studies, Plan International and Christian Aid, London, United Kingdom, 49pp.

Walz, A., Calonder, G.-P., Hagedorn, F., Lardelli, C., Lundström, C. and Stöckli, V. (2008) Regional CO_2 Budget, Countermeasures and Reduction Aims for the Davos Region, Switzerland. *Energy Policy*, 36, 811–820.

Wardle, D. A. (2003) Global Sale of Green Air Travel Supported Using Biodiesel. *Renewable and Sustainable Energy Reviews*, 7, 1–64.

Warren, R., Price, J., Fischlin, A., Santos, S. and Midgley, G. (2011) Increasing Impacts of Climate Change upon Ecosystems with Increasing Global Mean Temperature Rise. *Climatic Change*, 106(2), 141–177.

WeatherBill, Inc. (2008) Flight Disruptions and Weather: An Assessment of Weather's Effect on United States Airlines and Airports, San Francisco, CA, 58pp.

Weaver, D. (2011) Can Sustainable Tourism Survive Climate Change? *Journal of Sustainable Tourism*, 19(1), 5–15.

Webster, P.J. Holland, G.J., Curry, J. A. and Chang, H. R. (2005) Changes in Tropical Cyclone Number, Duration, and Intensity in a Warming Environment. *Science*, 309, 1184–1846.

Wikswo, M. E., Cortes, J., Hall, A. J., Vaughan, G., Howard, C., Gregoricus, N. and Cramer, E. H. (2011) Disease Transmission and Passenger Behaviors during a High Morbidity Norovirus Outbreak on a Cruise Ship, January 2009. *Journal of Clinical Infectious Diseases*, 52(9), 1116–1122.

Williams, G. and Geddis, G. (2010) Mitigation, Adaptation and Resilience: Linking Business Messages on Climate Change. Business in the Community, Newcastle upon Tyne, United Kingdom, 59pp.

Wilson, J. (2011) The Distribution of Climate and Weather Information for Tourism in New Zealand. LEaP Research Paper 3. Available at (18/04/2011) www.lincoln.ac.nz/leap

Wolfsegger, C., Gössling, S. and Scott, D. (2008) Climate Change Risk Appraisal in the Austrian Ski Industry. *Tourism Review International*, 12(1), 13–23.

Woodside, A. G. and King, R. I. (2001) An Updated Model of Travel and Tourism Purchase-consumption Systems. *Journal of Travel and Tourism Marketing*, 10(1), 3–26.

World Bank and the Queensland Reconstruction Authority (2011) Queensland Recovery and Reconstruction in the Aftermath of the 2010/2011 Flood Events and Cyclone Yasi. The International Bank for Reconstruction and Development, The World Bank Group, Washington, DC, and the Queensland Reconstruction Authority, City East Q, Australia, 80pp.

World Economic Forum (2009) Towards a Low Carbon Travel and Tourism Sector. Prepared with the support of Booz and Company. Available at (08/02/2012) www.theicarusfoundation.com/pdf/LowCarbonTravelTourism.pdf

World Heritage Convention Secretariat (2008) The Impacts of Climate Change on World Heritage Properties. Issues Related to the State of Conservation of Properties Inscribed on the World Heritage List. WHC-06/30.COM/7.1, World Heritage Convention Secretariat, UNESCO, Paris, France, 55pp.

World Meteorological Organisation (2011) WMO Statement on the Status of the Global Climate in 2010. World Meteorological Organization (WMO), Geneva, Switzerland, 20pp.

World Tourism Organisation (1998) *Handbook on Natural Disaster Reduction in Tourist Areas.* Madrid: WTO.

WTTC (2010) Climate Change. A Joint Approach to Addressing the Challenge. London, United Kingdom. Available at (20/11/2010) www.wttc.org

Wu, P. and Shi, P. (2011) An Estimation of Energy Consumption and CO_2 Emissions in Tourism Sector of China. *Journal of Geographical Sciences*, 21(4), 733–745.

WWF, Horwath HTL and HICAP (2010) Towards the Business Case for Sustainable Hotels in Asia. WWF, Horwath HTL and HICAP, 92pp.

Yaw, F. Jr (2005) Cleaner Technologies for Sustainable Tourism: Caribbean Case Studies. *Journal of Cleaner Production*, 13, 117–134.

Yeoman, I., Greenwod, C. and McMahon-Beattie, U. (2009) The Future of Scotland's International Tourism Markets. *Futures*, 41, 387–395.

Yu, G., Schwartz, Z. and Walsh, J. E. (2009) A Weather-resolving Index for Assessing the Impact of Climate Change on Tourism Related Climate Resources. *Climatic Change*, 95, 551–573.

Zacharioudaki, A. and Reeve, D. E. (2010) Modelling Shoreline Evolution in Response to Climate Change. *Proceedings of the International Conference on Coastal Engineering*, 32, 1–11.

Zeppel, H. (2011) Climate Change and Tourism in the Great Barrier Reef Marine Park. *Current Issues in Tourism*. Available online.

Zeppel, H. and Beaumont, N. (2011) Green Tourism Futures: Climate Change Responses by Australian Government Tourism Agencies. In 21st Annual Council for Australian University Tourism and Hospitality Education Conference (CAUTHE 2011): Tourism: Creating a Brilliant Blend, Adelaide, Australia, 8–11 February 2011. Available at (19/04/2011) http://eprints.usq.edu.au/18695/

Zientara, P. and Bohdanowicz, P. (2010) The Hospitality Sector. Corporate Social Responsibility and Climate Change. Chapter 5 (pp. 91–111). In C. Schott (Ed.), *Tourism and the Implications of Climate Change: Issues and Actions*. Bridging Tourism Theory and Practice. Volume 3. Bingley: Emerald.

Zotz, A. (2010) The Role of Tour Operators in Climate Change Mitigation. Nature Friends International and Respect Institute for Integrative Tourism and Development. Available at (23/08/2011) www.climalptour.eu/content/sites/default/files/Tour_Operators_EN.pdf

GEOGRAPHICAL INDEX

Abu Dhabi 146
Africa 3, 178, 200, 216
Antarctica 10, 18, 57, 211, 235,
Arctic 3, 10, 18, 28, 57, 65, 99–101
Australia 1, 13, 16, 19, 22–3, 25–27, 29, 42–9, 62, 83, 94, 112–3, 116–121, 145–7, 152, 162, 164, 170, 177, 182, 185–6, 189, 191, 201, 211, 224–225, 228–9, 234
Austria 20, 27, 123–4,

Bahamas 228
Bangladesh 199
Barbados, 20, 27, 103, 128–9
Belgium 58, 153
Bonaire 27, 144, 147

Cambodia 111
Canada 13, 18, 22, 24, 28, 47–9, 84, 100–1, 112, 142, 167, 187–8, 230,
Caribbean 13, 26–7, 29, 30, 36, 48, 53, 57, 72, 90, 95, 103, 124, 127–9, 141, 147, 175, 194–5, 222, 229
China 16, 18, 77
Cook Islands 74, 213–5
Costa Rica 144, 161

Denmark 20

Ecuador 159–160, 177
Egypt 22, 144
England 25, 27, 48, 49, 69, 94–98, 165, 177,
Europe 3, 38, 50–1, 58–61, 63, 68–70, 82, 104–5, 110, 111, 122, 124–6, 130, 136, 150, 152–3, 161, 182, 190,

Fiji 13, 86, 164, 185
Finland 27
France 40, 75, 106–7, 109–11, 160, 162, 221
French Polynesia 171, 208–210

Germany 17, 23, 40, 61–3, 97, 122, 124, 155, 158, 160, 175, 186–7, 233
Ghana 13
Great Britain 111, 130, 154, 171, 176,
Greenland 10, 99, 235

Hong Kong 22, 152

India 57–8, 175, 222
Indonesia 107
Ireland 58, 80, 124

Jamaica 38–9
Japan 17, 112, 155, 190

Maldives 58, 108, 196–197, 210
Malta 117
Mauritius 154
Mediterranean 23, 57–61, 72–3, 97, 130
Mexico 39, 41, 47, 49, 130, 147,
Middle East 130
Mozambique 223

Nepal 17, 52, 113, 114, 147,
Netherlands 16, 83, 108, 110, 153, 155, 160
New Zealand 13–7, 22, 25–6, 34–5, 87, 111, 123, 142, 147–8, 152–7, 159, 161, 170–2, 182, 185, 228, 234
North America 3, 20, 27, 48, 167, 178, 225
Norway 148

Oman 170, 234

Pakistan 3, 80–1, 225
Peru 177
Portugal 116

Russia 3

Saint Lucia 36–7, 103,
Samoa 41, 76
Scotland 41
Seychelles 18, 139, 144,
Singapore 153, 173
South Africa 18, 75, 116, 132, 170–1,226–8
Spain 23, 97–8, 117
Sri Lanka 144, 171
Sweden 12, 41, 143–4, 160, 171, 174
Switzerland 16, 23, 29, 186

Taiwan 17, 185
Tanzania 112, 199
Thailand 49, 148, 161, 174, 190
Tibet 52
Trinidad and Tobago 76, 103, 127, 182
Tunisia 40

Uganda 78–80, 211
United Kingdom 28, 40, 66–7, 104, 110, 119, 121, 168–9, 232
United States 13, 22, 27, 49, 61, 63, 69–71, 74–5, 90, 91, 95, 111, 122–3, 152, 163, 170–1, 173, 205–6

Vietnam, 161

Zambia, 13, 117, 216

GENERAL INDEX

Aall, C. 174, 223
Abdulla, A. 22
Abegg, B. 21, 26
accommodation 15–6, 19, 25, 48, 97–8, 140, 158, 163–7, 170–1, 177, 190, 207, 213–15, 231; *see also* hotel
Accor 106–107
accountability 36–7, 133,
accounting 15, 157, 174, 182, 185–6; accounting framework 184, 186
accreditation 107, 129, 147, 171, 190, 213
adaptation 31–3, 52–7, 101–2, 203–213, 215, 217, 235–9; action framework 3; and disaster risk reduction 223–29; *see also* disaster risk reduction; assessments 32, 33, 55–7, 61, 67, 71; barriers to 87–89; business level 41–49, 61–3, 82, 85–9, 238; defined 55; enabling environment for 31–4, 37–8, 44, 101; financing 34, 239; forced 55–6; local level 41–49, 78, 81, 83, 85; mainstreaming 38–9, 238; multi-country/international level 50–2; national plans 34–6, 78–81; national policies 35, 36; planned 55–6, 101; planning 31–56, 59, 62, 64, 67, 72, 77, 80, 84, 101, 237; plans 31–54, 62, 66, 82, 84, 237; policy 31–54, 101, 236–8; practice 55–102, 236–8; proactive: *see* adaptation planned; sub-national level 41–49; technologies 32, 34, 44, 75, 83
adaptive capacity 20–1, 32, 42, 53–4, 61, 84, 89, 119, 196, 199
ADB 64, 71
ADFC 157–158
advocacy 121, 134, 137–8, 215–17
Agrawala, S. 26
Aguilo, E. 117
air conditioning 25, 61, 67, 107, 110, 129, 157, 163, 171, 205, 207–8, 210, 215; *see also* deep sea water air conditioning system
air pollution 104, 113–14, 117, 137, 187, 213
air quality: *see* air pollution
air traffic management 135
aircraft 135, 149–155, 159; emissions 18–9, 172, 180; maintenance 155; operations 70; operators 125
airframe 151–2
airlines 18, 65, 68–70, 110, 120, 126–7, 133–6, 140, 151–5, 173, 178, 182, 185–6, 210; Airline Index 175; *see also* aviation
airport 68–9, 71, 95, 110, 111, 121, 124, 125, 135, 152–3, 171, 190
Al Marchohi, M. 229, 231, 233
Aleklett, K. 230
Allison, I. 10
alpine resort 23; *see also* skiing
Altamirano-Cabrera, J. C. 30
alternative fuels 134–5, 153; *see also* energy
Amat, J.P. 164
Amelung, B. 23
American Meteorological Society 73
Amr, A. 22
Andersen, P.A. 20
Anderson, K. 18, 150

Andrachuk, M. 19, 101
Anger, A. 126
Angus Reid 12
APEC 228
APEC International Centre for Sustainable Tourism (AICST) 26
Archer, J. 159
Arkell, B.P. 25
Arnell, N. W. 55
ASPIRE 133
Atmosfair 140, 175–6
Attractions: *see* tourist attractions
Australia's National Tourism Incident Response Plan 224
aviation 13–4, 17, 19, 120–1, 122, 125–6, 132–6, 140–1, 150–5, 178–9, 210, 230; *see also* airlines
awareness 1, 12, 22, 23, 26, 32, 37, 54, 90, 107, 132, 142, 163, 171, 172–4, 177–8, 185, 205, 211, 215, 219, 227
Aylen, J. 28

Baker, A. 210
Baker, C.J. 66
Baldock, T.E. 19, 164
Ballou, P. 211
Bank, M. 27
Bark, R. H. 62
Barnett, J. 57
Barr, S. 22
barriers 55, 62, 87–9
baseline 207, 220, 227
Bateman, I. 28
Bathols, J. 25
Bathols, J.M. 26
batteries 149, 152; *see also* energy
beach 13, 27, 147, 195, 213, 217; *see also* tourism beach
Beaumont, N. 13, 120
Becken, S. 1, 2, 4, 12–15, 17, 19, 22–4, 26, 27, 56, 57, 83, 86, 87, 95, 103, 105, 117, 118, 131, 137, 138, 149–160, 163, 164, 172, 228, 229, 231, 232, 233
behaviour: 14, 20–3, 29, 60, 97, 98, 176–7, 185, 211, 231; changes 141, 172–3, 199
beliefs 22
benchmark 119, 147, 171, 182, 187
Beniston, M. 26, 61
Bentley, R. 230
Benzie, M. 41
Berrang-Ford, L. 55, 56,101
Bett+Bike 158
Beukering, P. Van 170
Bicknell, S. 20
bicycle 115, 141, 147, 157–9
Bigano, A. 58
Bilimoria, K. 70
Bindoff, N.L. 10
Bindschadler, R.A. 10
biodiesel 18, 148, 168, 215; *see also* energy
biodiversity 27, 36, 39–40, 51, 72–5, 77, 81, 137, 159, 210
biofuels 230, 127, 135, 148, 150–4; *see also* energy
Bird, E. C. F. 75
Birkan, I. 23
Björnsson, H. 10
Blanco, J.M. 163
Bode, S. 162
Bohdanowicz, P. 104, 163, 190, 208
Bomhard, B. 77
Bonnefoy, P.A. 18, 150
Bonnell, R. 20
Borgström Hansson, C. 18
Bosello, F. 58
bottom-up 83, 186
Boucher, O. 18
Bows, A. 18–19, 150
Bradley, R. 205
brand 14, 104, 108, 142, 157, 187, 188, 198
Brander, L. 170
Brandt, A. 230
Brazel, A.J. 24
Bristow, A.L. 150
British Department of Transport 119
Broderick, J. 174
Brons, M. 173
Brouwer, R. 170
Bruno, P. 99
Buckley, C. 173
Buckley, R. 62
budget 143, 220, 232
building: design 188; standards 13
Building Energy Code 123
Burtt, M. 26–7
bus transport 141
bushfire 93, 225–6, 228
Butler, R. W. 95

Calbó, J. 59, 97
Calonder, G.P. 16
Campbell, J. 57
Campos-Soria, J.A. 23
Cancun Agreement 130–1
capacity building 78, 133, 200, 205, 213, 224
carbon: accounting 174, 181–191; calculator 116, 145, 160, 182; carbon-constrained 29; credits 145, 189; footprint 45, 140, 145, 147, 159, 162, 165, 171, 173, 174,

177–8, 182, 184–190, 236, 238–9; intensity 15, 18, 141, 150, 159, 174, 183; low carbon 110, 112, 126, 133, 173, 237; neutral 110, 113, 120, 129, 163, 144–5, 156, 159, 169, 170–1, 175, 178, 187–8; offsetting 14, 60, 110, 120, 141, 147, 154, 157, 169–173, 175–6, 178, 187, 189, 206–7, 236; sequestration 54, 103, 155, 176; tax 13, 46, 127, 163, 173, 233; trading scheme 46, 182, 211; *see* emissions trading
Cardenas, V. 24, 58
Caribbean Community and Common Markets 95
Caribbean Community Climate Change Centre 229
Caribbean Tourism Organization 90
car 15, 17–8, 111, 140, 147, 149, 157–9, 162, 165, 166, 182, 188,
carpooling 168
Cashman, A. 20
CAST 25–6, 94, 128, 213, 228, 229
Castley, J.G. 26–7
Cavan, G. 28
Ceron, J.P. 29,150, 173–4
Certified Emission Reductions 126, 129
Chan, W.W. 164
Chang, H. R. 223
Chapman, L. 66
Charara, N. 20
Chatterji, G. 70
Chen, H. 211
Chen, P. 185
Chen, Y.M. 164
Chester, M. 15
Cheung, C. 22, 172
Chitral Association for Mountain Area Tourism 81
Chmura, H. 77
Chomhairle Oidhreachta/the Heritage Council 80
Choy, Y.S. 16
Christensen, J.H. 10
Churie-Kallhauge, A. 162–3
Cioccio, L. 229
cirrus clouds 18
civil society 43, 54, 229
Cladera, A. 24, 163
Clapcott, R. 12–14, 117, 137
Clarke, R. 85
Clean Air – Cool Planet 174–5
Climate Commission 10
climate extremes 2, 8–9. 42, 64, 208, 224, 237
coastal tourism: *see* tourism coastal
coastal: low-lying coastal areas 12, 45, 57, 218; coastline 73, 80, 195
Coghlan, A. 24, 27
Cohen, S.A. 22
Colby, B. G. 62
Coles, T. 22
comfort 46, 58–61, 89, 215
Commission of the European Countries 51, 124–7, 144, 217
Commonwealth of Australia: *see* Government of Australia
communication 117, 136; activities 94; strategy 93–4,
community 12, 51, 78, 80–1, 84, 101, 133, 139–140, 143–4, 145, 163, 187, 221; community-based adaptation 196, 200
company policies 105, 137
competitiveness 51, 86, 89, 125, 126, 128, 144, 149, 174
conference 141, 155, 164, 216
consumer: attitudes 172; perceptions 163; *see also* behaviour
continuous descent approaches 151
contrails 18
Cook Islands News 58
Cooley, H.S. 185
Coombers, E.G. 27
coordination 12–3, 33, 41, 54, 94,
Copenhagen Climate Summit 132
coping capacity 32, 39; *see also* adaptation; *see also* adaptive capacity
corporate 142; responsibility 153; social responsibility 168, 169; customer 160; governance 163
corporations 31, 54
Cortes, J. 85
Cote I. 27
Coughlan, A. 27
Cowper-Smith, A. 18, 185
Cox, P.M. 10
Cramer, E. H. 85
crisis management 85, 93–94, 224–5, 228–9
cross-cutting 203–4
cruise ship 16–8, 28; 48, 65, 85, 91, 99–101, 141, 150, 159, 161, 211, 222, 236, 238
CSIRO 25
Curry, J. A. 223
cycle tourist 141, 157–9; *see also* bicycle
cyclone 3, 8, 43, 45, 65, 83, 86, 89, 92–3, 223–5

Dalton, G.J. 19, 164
Darch, G.J.C. 25
Davidson, D. 91
Davos Declaration 15, 53, 131, 145, 216

Dawson, J. 18, 22, 27, 57, 61, 62, 101
de Bruijn, K. 16
de Freitas, C. 27, 87
de Grosbois, D. 19, 185
de Noblet, N. 10
decision making 5, 12, 37, 64, 66–7, 78, 81, 83, 89, 98, 188, 216, 220, 231, 236
decision support 32, 34
deep ocean water-cooling technology: *see* deep sea water air conditioning system
deep sea water air conditioning system 208–10
DEFRA 147
DeLacy, T. 21, 55, 58
Department of Resources, Energy and Tourism 118, 211
departure tax 122–4
Des Bois, R. 100
desalination 147
destination 236–9; attractiveness/appeal/choice 5, 51, 56, 57–63, 81, 83–4, 139, 141, 211, 231–233; attribute, 58; experience 56–7; last chance 57, 80; marketing 140, 142, 185; safety 81–85, 89–92; sustainability 57, 87–9, 95–102, 205, 213, 216–22, 224, 226–8; risk 57, 226–9; water availability/consumption 73, 97–8, 214–16, 221–22, 226–8
development project 114, 218
Dhakal, S. 113–4
Dickinson, J. 15, 19–20, 158, 170
Dickson, T. 13, 20
Dieke 33
Dirven, R. 16
disaster 40, 84, 93, 214, 224–5, 228–9; management 84–5, 95, 214, 224–5, 228–9, 237; natural 3–4, 26, 93, 216; risk management 31, 195, 201, 204, 223–9; risk reduction 31–3, 49, 82, 204, 212–13, 223–9, 237–8
district heating 187
Djerba Declaration 103, 131
Dobney, K. 66
Dodds, R. 13, 116, 142–3
Dokken, D. 151
Dolnicar, S. 22
domestic tourism 1, 16, 40
Dominguez, F. 62
donors 193
Dorfman, M. 95
Draper, D. 99, 101
Draper, S. 19, 28, 232
driver education 110
Dubois, G. 29, 150, 174
Duerden, F. 101
Duffy-Mayers, L. 128–9, 164
Duijnisveld, M. 15
duty 115, 120–2
Duvat, V. 217, 219
Dworak, T. 41
Dwyer, L. 16, 173, 186
Eang, L.S. 163
early warning 41, 57, 66, 84, 89–92, 205
EarthCheck 119–120, 147, 171, 182, 208

ecolabel 140, 190
ecological footprinting 15
economic 36, 40, 43, 54, 61–2, 64, 93, 203, 211, 217–8, 231, 234, 238–9; growth 12, 143; incentives 36; impacts 29, 46, 61, 118, 145; 152; measures 134–5; opportunities/benefits 44, 46, 205, 212, 216, 226; resilience 43; risk 62; sectors 39, 53, 229
ecosystems: *see* natural systems/resources
education 34, 37, 54, 78, 83, 213–14, 222
Ehmer, P. 23, 28
Eijgelaar, E. 16, 18, 178, 211
Eischeid, J. 73
elasticities 27, 210
electric vehicle 114
Elliott, V. 173
emission: factors 183; reduction targets 104, 110, 237; calculator 116, 182, 190; targets 104; trading 211, 234
Emission Reduction Units 126
Emission Trading Scheme 121, 125–6, 157, 173, 182, 233
emissions-based taxes 210
Endler, C. 23
energy: audit 15, 19, 124, 128, 143, 215, 236; conservation 205, 215–16, 236–7; efficiency 205; renewable energy 16–9, 104, 107, 125, 127–9, 130, 143, 147, 164, 166, 167, 170–1, 182, 187, 190, 195, 205–6, 210, 216; saving 17, 236; security 204, 206, 213, 222, 230, 232–4; system 163, 188
EnergyStar 162–3
enforcement 13, 37, 86, 114, 192
Engelbrecht, C. J. 226
Engelbrecht, F. A. 226
England, M.H. 10
environmental: awareness 22, 185, 211; management 66, 104, 106, 116, 134, 137, 163, 196, 215, 236
erosion 36, 42, 57, 75, 86–7, 218
ethical 19, 83
Eugenio-Martin, J.L. 23
EUROCONTROL 68, 153

European Commission: *see* Commission of the European Countries
European Community 152
evaluation 32–4, 44–5, 102, 180, 191–2, 217–18, 236, 238
events organisers 96, 164, 169

Fahey, D.W. 150
Falk, M. 26
FAO 85
Faulkner, B. 229
feedback 188, 203
Fenner, R. 26
Ferris, E. 3
Filimonau, V. 15
finance 32, 206, 215, 220, 234, 238;
financial: savings 210; opportunities: *see* economic opportunities; problems 220; resources 180; tools/instruments 126;
financing/funding 32, 44, 89, 209–10, 213, 218, 220, 226, 239
Firestone D. 25
Fischer, E.M. 59
Fischlin, A. 75
flood 3, 82, 96
food 57, 214–15, 222–3; safety 85, 91–92; security 57, 213
Forbes, D. 99
Ford, J. 55, 56, 101
forestry 143
Forster, P.M. 150
Forsyth, P. 16, 173, 186
fossil fuels 12, 103, 143, 153, 159, 184, 205, 207, 215; *see also* energy; *see also* fuel
Fourth Assessment Report 10, 235; *see also* IPCC (Intergovernmental Panel on Climate Change)
Frampton, C. 163
Francis, J.E. 10
Frank, J. M. 85
fuel: surcharges 173; consumption 150–7, 178, 211, 215; combustion 149; substitution 215; *see also* energy; *see also* fossil fuels
funding 14, 89, 115, 118, 143–4, 210, 220

Gardiner L. 25
Garrod, B. 174, 223
Geddis, G. 204
Gehr, R. 20
geothermal 147, 155, 170
GHG: *see* greenhouse gases
Gill, J.A. 27–8
Gittens, A. 72
Gladin, E. 53, 56, 114
Glastonbury Festival 95–7
Gleick, P.H. 185
Glick, P. 77
global agreement 126, 130
Global Travel 173
Glynn, S. 28
golf 28, 42, 82, 97, 171, 226
Goodman, J. 28, 232
Gordijn, H. 122
Gössling, S. 1, 4, 12–3, 18, 20, 27–9, 56, 104, 136, 144–5, 150, 174–5, 194, 223, 237
government 31, 36, 43, 44, 54, 84–5, 93, 216–17, 220, 234, 236, 238
Government of Australia 45, 46, 225
Government of France 210
Government of French Polynesia 210
Government of Jamaica 39
Government of Saint Lucia 36
Government of the United Kingdom 66, 172, 119–121
Goyette, S. 26, 61
Grabbe, S. 70
Graci, S. 13
Great Barrier Reef 27, 48, 162, 193, 237
Green Globe 147, 213
green-washing 146
Greene, D. L. 230
Greenhouse Gas Protocol 185–6
greenhouse gas 15–18, 48, 51, 54, 57, 69, 103–179, 185, 188, 205, 208–9, 211, 214, 216–17, 221, 223, 233–4, 236–8
Greenland Ice Sheet 10, 99, 235
Greenwod, C. 28
Gregoricus, N. 85
Griggs, D. 151
Gruber, N. 10
Guihua, Y. 18

Hadwen, W. 13, 20, 25
Hagedorn, F. 16
Hall, A. J. 85
Hall, M.C. 12, 28, 53, 56, 114
Hallegatte, S. 95
Hamilton, J.M. 58
Hammill, A. 205
Handley, J.F. 28
Hänsel, S. 61, 62
Hansman, R.J. 18, 150
Hapke, J. 162
Harding, P. 164
Hardyment, R. 28, 232
Hares, A. 170
Hartz, D.A. 24
Hatate, Y. 164
Hay, J.E. 4, 17, 57, 95, 104–105, 139, 149

Haywood, A.M. 10
Heath, E.T. 18
heating 16, 143, 163, 166, 187
heatwave 3, 9, 24, 58–61, 72, 84, 97, 207–8, 225
Hein, L. 23
Heisler, G.M. 24
heliski operator 88
Hendrikx, J. 26, 83, 87
Hennessy, K. 25–6
heritage 57, 75, 77–8, 82, 216–17, 236; sites 45, 51–52, 75, 77–8; tourism 80
Hermanus Water Resource Development and Management Programme 226
Heymann, E. 23, 28
HICAP 208
Higham, J.E.S. 22
Hihara, K. 18
Hille, J. 174, 223
Hills, J. 164
Hirsch, R. L. 230
HM Revenue and Customs 120–4
Hoerling, M. 73
Hoffman, A. 21, 26
Hoffman, V.H. 12
Holecek, D. 13, 27
Holland, G. J. 223
Honnery, D. 149–150
Höök, M. 230
Hoolhorst A. 17, 151
Hopson, J. L. 230
Hoque 16, 186
Horner, J.D. 211
Hörstmeier, O. 18
Horvath, A. 15
Horvitz, A. H. 68
Horwath H. T. L. 208
hot water 17, 107, 127–9, 163–4, 170, 215; *see also* water
hotel 13, 16, 20, 36, 48, 75, 86, 97–8, 106–8, 123, 127–8, 130, 147, 163–4, 171–2, 182, 106–7, 190–1, 207–9, 213–215, 222, 236; *see also* accommodation
Hou, G.Q. 164
Houghton, M. 225, 228
Houston, J.R. 75
Houston, T.G. 68
Howard, C. 85
Howell, S.E.L. 99, 101
Howitt, O. 159
Hoy, A. 61, 62
Hreinsson, E. 26
Hudson, S. 84
Huebner, A. 57
Hughes, L. 229, 230, 234
Hughey, K. 83, 87, 228
Hunter, C. 27, 212
hurricane 3, 9, 36, 39, 43, 45, 47–9, 53, 63, 65, 83, 86, 89–93, 223–4, 226, 228; policy 46
Hutchinson, M. 26
hybrid 17–8, 149, 159, 168, 170
hydropower 114, 127,
Hyogo Framework for Action 226

IATA 119, 134–6, 150,-3, 173, 234
IFC 71
incentive 5, 91, 104, 118, 129, 133, 153, 168, 178, 192; to travel 57–63, 232
income 75, 211, 221, 223
indicators 25, 27, 33, 40, 107, 108, 192–5, 218, 236
industry/tourism organizations 31, 37, 41–2, 51, 53, 54, 84, 222, 224–6, 235, 237–8
information sharing 34, 123,
infrastructure 36, 39, 45, 47, 50–1, 53, 56, 64–72, 79–80, 82, 89, 93, 96, 100, 221, 224, 226–7, 236
injuries 194
innovation 4, 20, 41, 108, 133, 141, 143, 208, 228; innovative 27, 86, 141
Instinct and Reason 172
institutions/institutional arrangements 31–4, 54, 59, 78, 101, 216–17, 236, 238
instrument 117–9, 122, 125, 129, 130, 134, 182, 192,
insurance 61, 63
integrated coastal zone management 75, 217–21
integrated responses 33, 37, 48, 50, 53, 203–4, 211–21, 224, 226, 229, 235, 237–9; *see also* synergies
Intercontinental Resort and Thalasso Spa 208
International Air Passenger Adaptation Levy 13, 210
International Air Passenger Levy 208
International Energy Agency 230
International Union of Railways 25, 50
inventories 183, 185, 187, 189
investment 13, 20, 27, 29, 119, 127, 129, 133–4, 152, 156, 188, 196, 210, 230
IPCC (Intergovernmental Panel on Climate Change) 3, 8, 10–1, 42, 57, 66, 99, 104, 151, 169, 118, 149, 172, 182–3, 207, 223, 235

Jabbour, J. 223
Jakobsson, K. 230
Jamaica Tourism Task Force 39

Jariabka, M. 100
Jatropha 153–4
Jaykus, L.A. 85
Jeuring, J. 23–4
Jiminy Peak Mountain Resort 206–7
Jones, A. 4, 27–8
Jones, B. 24, 61–2
Jones, S. 78, 211
Jopp, R. 21, 55, 58
Jorgensen, B. 13, 20, 25

Kai, T. 164
Kalash 81
Karoly, D.J. 10
Kaser, G. 10
Kavoussi, S. 69
Kean, I. 84
Keesman, A.T. 163
Kelman, I. 116
Keskin, B. 163
King, R.I. 231
Klein, A. 69
knowledge sharing 199, 201, 218
Koehler, J. 126
Kolb, M. 21, 26
Konopek, J.
Kozak, N. 23
Krarup, S. 138
Kumar, S. 163
Kummer, B. 91
Kuo, N. 185
Kwon, T.H. 150
Kyoto Protocol 104, 112, 130

Laaser, C. 41
labelling 166, 174, 190
Laesser, C. 22
Lai, T.M. 16
Laing J. 225, 228
Lam, J. C. 164
Lam, K.H. 16
Lambert, E. 27
Lardelli, C. 16
Lassen, C. 20
Law, R. 22, 172
Le Quéré, C. 10
leadership 20, 81, 106–7, 132, 138, 143, 159, 167, 216, 234
learning 45, 101, 182, 199–202, 220, 238–9; experience 141; opportunities 182;
Lee, D.S. 150
Lee, J.J. 18, 150
Lee, R. S. 69
Lee, S. 19
legal obligation 192
legislation 32, 34, 66, 85, 88, 91, 94
Lehmann, L.V. 73
Leigh, C. 232
Lemelin, H. 18, 22, 57
Lemieux, C. 24
length of stay 16, 18
Lenton, T.M. 10
Li, D. 164
Li, J. 230
life cycle 15, 95, 153, 163, 169, 174, 185
light bulbs 107, 141, 166, 168, 207
Lim, L.L. 150
Lin, T. 17
Lindley, S. 28
Linham, M. 75, 83
Lister, D. 151
Lo, W.C. 16
Lobo, A. 116
Local Agenda 145
local: government 14, 31, 41, 104, 112, 139, 144, 220; locally grown food 143, 170, 173; locally-owned 169; participation 112
Lockington, D.A. 19, 164
Lonely Planet 172
long distance 66, 150, 179
Lopez-Morenob, J.I. 61
Losenno, C. 41
low-cost carriers 28, 113
Lucas, C. 25
Lueck, M. 57
Lumsdon, L. 19–20, 158
Lundström, C. 16
Lusaka Declaration 215–16
luxury 91, 108, 141, 159,
Lynes, J. 19

Macintosh, A. 17, 150
MacLeod, C. 27
Maher, P. 57
Maibach, M. 15
mainstreaming 11, 38–9, 123, 283
Mair, J. 21, 22, 55, 58
Mak, L.M. 164
maladaptation 98, 198–9, 203
Malone, R. 164
Management Consultancies Association 106
Mann, M.E. 10
marketing 12, 63, 83, 85–7, 98, 116, 119, 129, 138, 140, 178, 185, 225–6
Marshall, N. 22
Marshall, P. 22
Martinac, I. 162–3, 208
Martinez, V. 24, 163
Mas, F.J. 172

Massachusetts Green Communities Act 206
Massachusetts Technology Collaborative 206
Massey, E. 41
Matasci, C. 30
Matschullat, J. 61, 62
Matson, J. 18
Matus, K. 22
Matzarakis, A. 23
May, A.D. 150
Mayes, G. 87
Mayor, K. 13, 150, 173
McBoyle, G. 26–7, 61–2
McEvoy, D. 28
McFarland, M. 151
McGray, H. 205
McGregor, J.L. 226
McGuirk, M. 68
McKercher, B. 22, 172
McMahon-Beattie, U. 28
McManus, P. 20
McMorrow, J. 28
McMullen, C.P. 223
McNamara, K. 24
McNeil, B.I. 10
McQuatters-Gollop, A. 85
media 11–2, 14, 90, 96, 99, 134, 143, 172, 179, 216–17, 225
Mehta, M. 95
Mensah, I. 13
Met Office: *see* UK Meteorological Office
Metzger, M. 23
Michael, E. J. 229
Michalena, E. 164
micro hydro plant 128, 147, 167, 170; *see also* energy
Middel, J. 17, 151
Midgley, G. 75, 77
Millennium Development Goals 53, 216
Miller, G. A. 84
Miller, R. 230
Mills, B. 26
Mills, G. 25
Minogue, A. 26
Mitchell, T. 212
mitigation 6, 13, 15, 19, 22–3, 31, 34, 39, 42, 45–6, 56, 103–97, 201–11, 215, 217, 232–3, 235–9; targets 182
mobility 20, 28–9, 123, 149, 160
model/modeling 51, 58–9, 69, 75, 90, 95, 97, 231
Moia, A. 24, 163
monitoring 32–4, 36–7, 44, 46–7, 51, 78, 99, 102, 217–18, 238
Monroe County (Florida Keys) Tourist Development Council 90–1
Moore, S. 84
Moreno, A. 23, 27, 58
Moriarty, P. 149–150
Morrell, P. 18, 151
Müller, H. 29
multi-criteria analysis 182
Murray, V. 19, 28, 232

NASA 70, 71
National Hurricane Centre/Tropical Prediction Centre 47, 90, 92
national park 17, 24–5, 45–6, 52, 78–9, 141, 166, 175, 211, 228
National Ski Areas Association 203
Natori, R. 164
natural systems/resources 36, 39, 45, 51, 53–4, 55–6, 72–3, 77, 81, 212, 209, 212, 216–18, 221, 222–3, 226
Needham, M. 22
Nelson, V. 23
Nepal, S.K. 17
Network Rail 66, 67
New Zealand Ministry of Tourism 34, 117
Newquay Zoo 98
Newton, P.J. 150
Nicholls, R. J. 75, 83
Nicholls, S. 13, 23, 27
Nicolau, J. L. 172
Nijkamp, P. 173
NOAA 2, 11
Noland, R. B. 69
Nolitours 47
Nolte, R. 50
non-governmental 5, 36, 222
Nygren, E. 230

OECD 12, 32, 33, 50, 117, 149, 199, 205, 212, 236
Office of Rail Regulation 66
offsetting: *see* carbon offsetting
oil prices 28, 113, 150, 173, 178, 230–4
Oléron Coastal Quality Action Plan 217
opportunities 28, 40–1, 44, 46–8, 51–4, 55–7, 62, 82, 86, 95, 99–101, 141, 158, 165, 169, 173. 182, 191, 204–5, 214, 222, 229, 232, 239; *see also* economic opportunities
Orchiston, C. 224
Otto, I. 18
Overstrand Municipality 226
Owen, B. 150

Pacific Resort Hotel Group 213–15
package tourism 106, 140

Pambudi, D. 16
Park, S. 19
Parkman, M. 72
passenger 16, 17, 25, 48, 85, 91, 100, 115, 120–1, 134, 150–6, 159, 161, 169, 173
Patterson, J. 55, 56
Patterson, M. 186
peace 215
peak oil 43, 229–34
Pearce, T. 101
Peeters, P. 13, 16, 18–9, 18, 104, 136, 150, 117, 174, 178, 210, 223, 237
Pegion, P. 73
Pejovic, T. 69, 70, 211
Pels, E. 173
Peng, L. 18
Penner, J. 151
Pentelow, L. 2, 13
perception 23–4, 34, 59, 96, 163, 172–3, 185, 188, 236
Perch-Nielsen, S. 16, 27, 186
performance 66–7, 71–2, 238–9; assessment 180; economic 40; evaluation 40, 217; measurement 33, 218; operational 51, 216; targets/indicators 33–4, 44–5, 66, 218
Perlwitz, J. 73
permits: *see* emissions
Peterson, T. C. 68
Petz, D. 3
Pham, T. 13, 20, 29
Phillips, M. 4
photovoltaic electricity 127–8, 144, 164, 167, 170; *see also* energy
physical planning 36–7, 39
Pickering, C. M. 26–7, 62
Pierce, G.J. 27
Pitman, A.J. 10
planning: initiatives 50, 181; instrument 35, 192, strategies 188
polar bear 18, 22, 57
policy: gaps 13–4; policy instruments 18, 129, 182; making 13, 35, 37, 41, 50, 54, 112, 117, 122, 137, 138, 192, 200
Price, J. 75
Prideaux, B. 22, 24, 27, 95, 172
Pridmore, A. 150
Prillwitz, J. 22
private sector 33, 36, 43, 54
private-sector investments 196–7
Priyadarsini, R. 163
projections 59, 61–2, 68, 224, 235
psychological 20, 149
public transport 114, 140, 147, 149, 156, 158, 165, 169, 218
Quan, X. W. 73
Queensland Reconstruction Authority 225
Queensland Tourism Crisis Management Plan 93
Quinn, A. 66
Quiros, K. 163

Rahmstorf, S. 10
Rail Safety and Standards Board 66, 71
Räisänen, J. 10
Ramesohl, S. 138
Ramírez Sánchez, P. 18
Rao, P 173
Rebetez, M. 23–4
recycling 95, 207, 214–16
Reddy, V. 15
Reeve, D. E. 75
refrigeration 107, 110, 123, 163, 166, 209
regulations 5, 34, 66, 81, 85, 91, 98–100, 114, 118, 163, 220
Reisinger, A. 24, 231
Rendeiro Martín-Cejas, R. 18
renewable energy: *see* energy
rental cars 111, 140, 159
reporting 32–3, 36, 44–5, 55–56, 66, 78, 215–16, 225
research 34, 37, 54, 55–6, 78, 82, 210, 215, 218, 225, 232, 235–6, 238
Research and Development 118
resilience 4, 13, 32, 43–6, 49–53, 64, 66, 71, 82, 95, 98–9, 192, 196–9, 212, 221, 224, 229, 235
resource base 40, 49, 56, 72–8, 81–3
responsible tourism 83, 111, 116, 142, 145, 177, 216, 222
Revol, V.G.N. 159
Ribeiro, M.M. 41
Richardson, K. 223
Riera-Font, A. 24, 58
Riera, A. 117
Rietveld, P. 173
Rignot, E. 10
risk 33–4, 41–2, 44, 99, 207, 224; assessment 32, 44, 47, 49, 51, 55, 57; business 31, 54, 62, 82; climate 31, 33, 35–41, 44, 50–1, 54–5, 83, 84, 101, 239; destination 42–45, 47, 53–4, 57; drought 51, 226; health 59, 85, 91–92; hurricane/cyclone 46, 83; infrastructure 47, 66–7, 95; injury 46; management 49, 50, 53–4, 55, 84, 237; reduction 31–2, 39, 51, 238; tourism 34, 49, 61
Ritchie, B. W. 224
Robbins, D. 15
Robertson, D. 84

Rodger, C.J. 159
Roson, R. 58
Rossello-Batle, B. 24, 163
Rossello-Nadal, J. 24, 58
Rossello, J. 117
Rossetti, M. 66
Rough Guide 177
Rouphael, T. 22
Royal Academy of Engineering 64
Rudolph, J. 229, 230, 234
Rummukainen, M. 10
Rutty, M. 23, 59

Saggel, S. 18
Sanders, D. 225, 228
Santamarta, L. 23
Santos, S. 75
satisfaction 27, 46, 83, 86–7, 94, 109, 142, 185
Sausen, R. 150
scenarios: scenario analysis 106, 188, 236
sceptics 11
Schär, C. 59
Schellnhuber, H.J. 10
Schernewski, G. 28
Schiermeier, Q. 11
Schiff, A. 231
Schillemans, R.A.A. 163
Schneider, C. 15
Schneider, S.H. 10
Schuhmacher, K. 145
Schwartz, Z. 23
scope 184
Scott, D. 13, 18, 23–4, 20, 26–8, 56, 59–62, 72, 95, 104, 131, 136, 175, 178, 194, 237
sea level 3–4, 8, 10–11, 27, 29
season 23–4, 28, 45, 48, 59, 61–65. 69, 73, 84, 90, 99, 187, 205, 217, 226
security 215; *see also* energy security
segmentation: *see* market segments
Serquet, G. 23–4
Sesartic, A. 16
Sgouridis, S. 18, 150
Sharples, J. 26
Shaw, G. 22
Sherwood, S.C. 10
Sheth, K. 70
Shi, P. 16
Shih, C. 27
Showler, D. 28
Simmons, D. 13, 20, 25, 29, 163
Simpson, M.C. 53, 56, 72, 95, 114, 175, 194
Sisman and Associates 18
ski tourism: *see* ski; *see* skiing
skiing 27, 167–8, 187–8; cross country;
ski: destination 61–2; fields 21, 27; industry 20, 26; resort 20, 59, 141, 176, 193–4, 205–6; slope 207
slow tourism 20, 157–8, 236
Smerecnik, K.R. 20
Smit, F. 101
Smith, I.J. 159
Smith, I.N. 26
Smith, T. 87, 101
snow cover/duration 49, 59, 61–3, 79, 203
snowmaking 27, 62, 205, 207
snowmobiling 27, 61, 168
snowstorms 3
SNV Netherlands Development Organisation 83
social responsibility: *see* corporate
Söderbergh, B. 230
solar: control window film 163; energy 145, 170; hot water 17, 107, 123–4, 127, 164, 170, 215; lamp 127; mirrors 175; panels 143, 159–160; *see also* energy
Somerville, R.C.J. 10
Sorrell, S. 230
space tourism 18
Speirs, J. 230
Sprengel, D.C. 21, 26
Spurr, R. 16, 29, 186
Sridhar, B. 70
Statistics Greenland 99
Steele, P. 135
Steffen, K. 10
Steig, E.J. 10
Stein, B. A. 77
Stewart, E.J. 57, 99, 101
Stewart, E.J. 18, 22
Stöckli, V. 16
storms: surges 8, 36
Strasdas, W. 174
strategy 136, 145, 168, 191–5, 199, 227, 237; *see also* planning
Stucki, M. 16
subsidies 118, 182, 210
Sugden, R. 28
summer 59–62, 68–9, 81, 96, 99, 235; hotter 41, 48–9, 59–61, 68, 84
Sunshine Coast Council 42–5, 112–3, 234
supply chain 15, 105, 107, 110, 133, 146, 169, 184, 189, 222–3, 224
sustainability 5, 19, 30, 57, 204–5, 208, 211–23, 226–8, 238; *see also* destination sustainability
sustainable development: *see* sustainability
sustainable: business 111, 133
Sutherland, W. 27–8
Swart, R. 41

swimming pools 86, 97, 107, 163
synergies 32–3, 119, 203–10
Szimba, E. 15

Takahashi, T. 164
tax 75, 115, 117–120, 124, 126–9, 173, 180, 206, 210, 233; *see also* carbon
taxi 160–1
technology 17, 44, 89, 108, 118, 125, 133–5, 141, 149, 151–3, 163, 168, 187, 199, 205–10, 212–16, 230, 239
Tervo, K. 27
Thaper, C 18, 178, 209
The Greenhouse Gas Protocol Initiative 184
The International Ecotourism Society 136, 159, 202
The International Tourism Partnership 4, 184, 222, 239
third generation biofuels 149, 153; *see also* energy
Thomsen, D. 87
Tight, M. 150
Tinch, R. 27
tipping point 22
Tirado, D. 73, 97
Tirado, M. C. 85
Tivy, A. 99, 101
To, W.M. 16
Tol, R.S.J. 13, 58, 150, 173
top-down 183, 186
tornadoes 26, 94,
Tortella, D. 73, 97
Toumi, R. 69
tour operators 6, 105, 136, 139–141, 168–9, 175
Tourism Australia 22, 185
Tourism New Zealand 142, 172, 185
tourism operators 31, 33, 41, 50, 54–5, 62, 83–4, 87–8, 93–4, 101, 203, 205, 211, 213, 222, 228–9, 231, 237
Tourism Queensland 26, 93, 177, 225
Tourism Research Australia 83
Tourism Victoria 26, 116, 120, 229
tourism: coastal 27, 36–7, 40, 42–45, 53, 58–60, 63, 65, 71, 73–6, 78, 94, 217–21, 224, 236; heritage 80, industry 72, 83–4, 89, 90, 93, 205, 208, 216, 222, 225, 228–9, 232; last chance 46, 57, 80; management 239; mountain area 61–3; 78–81; planning 24, 29, 148; sustainable 39, 48, 51, 211–16; 221–22, 232; system 54, 58, 101; ski 40, 48, 59, 61–63, 88, 205–7, 236
Tourism Satellite Account 186
tourist 31, 33, 34–5, 54–5, 57–9, 61, 75, 81–5, 89–91, 94, 97, 101, 211, 217, 231, 236; attractions 36, 72, 98, 139, 140, 144, 171, 217; health and safety 81–94, 224–5; season: changes in 45, 48–9, 57–9, 62, 101, 205
traffic management 135
train 16–7, 25, 66–8, 110, 150, 160, 176, 232
training 37, 54, 78, 84
transformations 150, 216
transport 56, 64–72, 88, 91–2, 99–101, 211, 216, 218, 224, 231–2; mode 15–19, 141, 149–150, 173231–2
transportation 15, 17, 117, 136, 146, 149–162, 178, 185, 187–8; *see also* infrastructure
Transportation Research Board 64, 65, 66
Tratalos, J.A. 28
Tremblay, P. 13, 20
tropical cyclones: *see* hurricanes
Trotz, U. 72, 95
Trung, D.N. 163
Turbine: *see* wind turbine
Turton, S. 13, 20, 25
typhoons: *see* hurricanes

U.S. Federal Food, Drug, and Cosmetic Act 91
U.S. Food and Drug Administration 91
U.S. Food Code 91
U.S. Passenger Vessel Services Act 91
U.S. Public Health Service Act 91
Uemura, Y. 164
Uhlmann, B. 26, 61
UK Climate Change Act 66
UK Energy Research Institute 230
UK Meteorological Office 67
UK Parliamentary Office of Science and Technology 72
uncertainty 11, 20, 52, 62, 191, 227
UNCTAD 99
UNDP 210
UNEP 2, 12–3, 15, 20, 25, 28, 32, 33, 50, 57, 63, 68, 75, 112, 117, 127, 130, 140, 149, 168, 171, 177, 100, 212, 213, 216, 226, 227, 228, 236
UNESCO 28, 81, 168
UNESCO World Heritage Centre 77
UNFCCC 130, 132, 175, 200, 215
UNISDR 226
UNWTO 1–2, 13, 15, 20, 28, 57, 61, 75, 130–2, 138, 145, 149, 168, 194, 199, 212, 216, 228, 238
Upham, P. 4, 174
Urban, F. 212

URS 64
US Energy Information Administration 230
USAID 114
USEPA 72
USFDA 91
Uyarra M. 27
Uysal, M. 23

values 22, 45, 51, 80, 81, 157, 217–18
van Egmond, T. 2
van Ho, T. 16
Vaughan, G. 85
Verbruggen, A. 229, 231, 233
vessels 71, 91–2, 100–1, 156–7, 159–160, 211
Viner, D. 23, 27
Visbeck, M. 10
Visser, N. 2
voluntary 27, 138, 145, 156, 178, 192; agreements 118; initiatives 104–5;
volunteer tourists 177
vulnerability assessment 51, 55

Waibl, G. 25
walking 17, 80, 114, 145, 149
Wallace, L. 17, 150
Walsh, J.E. 23
Walsh, K. 26
Walz, A. 16
Wang, Y.H. 164
Wardle, D. A. 18
Warren, R. 75
water 36–7, 39–41, 44–5, 48, 51, 57, 62, 72–3, 77, 82, 91, 95–9, 214–16, 221, 222, 224, 226–8, 236, 239; use 20, 97–9, 226–8; *see also* hot water
Watkinson, A. 27–8
weather 58–9, 61, 64–5, 67–9, 72, 81, 83–4, 87–91, 95, 101, 215; extreme 25, 66–8, 70, 77–8, 82, 85, 89–91, 95–97, 208, 223–4, 237; information 24–5, 87; surprises 223; warnings 23, 51
WeatherBill Inc 69
Weaver, A.J. 10
Weaver, D. 22, 25, 134
Weber, F. 29
Webster, P. J. 223
Wehner, M. S. 68
Whetton, P.H. 26
WHO 85
Wiesner, R. 27
Wikswo, M.E. 85
Wildlife 28, 48, 101, 154, 167, 188, 223
Wilkes, K. 170
Williams, G. 204
Williams, V. A. 69
Wilson, J. 24–25, 35, 83, 87, 231
Wilson, R. 13, 20, 25
wind energy 164, 174, 205; *see also* energy
wind turbine 205–7
wingtips 151
Wit, R.C.N. 150
WMO 2, 8, 11, 13, 15, 20, 28, 57, 63, 68, 149, 212
Wolfsegger, C. 20
Woodside, A. G. 231
World Bank 210, 225
World Economic Forum 19, 104
World Heritage Convention 77
World Heritage Convention Secretariat 51, 78
World Travel & Tourism Council: *see* WTTC
WRI 213
WTTC 4, 104, 107, 116, 132–4, 184, 238–9
Wu, P. 16
WWF 208, 210

Xie, H.R. 164
Xuchao, W. 163

Yackel, J. 99
Yaw, F. Jr 124
Yeoman, I. 28
Yoshida, M. 164
Yu, G. 23

Zacharioudaki, A. 75
Zeppel, H. 13, 48, 120, 161, 170
zero-carbon 146, 162, 182; *see also* carbon
Zhang, T. 73
Ziegler, A. 21, 26
Zientara, P. 104, 190
Zisler, S. 162
Zotz, A. 105, 168